Synthesis Lectures on Computer Science

The series publishes short books on general computer science topics that will appeal to advanced students, researchers, and practitioners in a variety of areas within computer science.

Hang Luo

Influence Models in Group Decision-Making

 Springer

Hang Luo
Peking University
Beijing, China

ISSN 1932-1228   ISSN 1932-1686 (electronic)
Synthesis Lectures on Computer Science
ISBN 978-3-032-01351-4   ISBN 978-3-032-01352-1 (eBook)
https://doi.org/10.1007/978-3-032-01352-1

This Springer imprint is published by the registered company Springer Nature Switzerland AG
The registered company address is: Gewerbestrasse 11, 6330 Cham, Switzerland

If disposing of this product, please recycle the paper.

To my mentors and family.

Preface

It is quite common that people influence and are influenced by each other in group decision-making. Likewise, artificial intelligences can influence and be influenced by each other in interaction or collaboration. Nowadays, both a person and an artificial intelligence can be called an agent. In this book, we consider settings of group decision-making where agents' preferences/choices are influenced (and thus changed) by each other. As the influence of reality faced by an agent usually comes from more than one agent (at the same time), we provide both cardinal and ordinal approaches to address multiple sources of influence in group decision-making, particularly considering varied strengths (stronger or weaker) of different influences and opposite polarities (positive or negative) of different influences. We extend classical social choice functions, such as Borda count and Condorcet method, to signed and weighted *social influence functions*. More importantly, we extend the KSB distance metric to a *matrix influence function*: we define the rule of transforming every preference ordering (over alternatives) into a corresponding matrix (named the ordering matrix) and set a metric to support the computation of the distance between any two ordering matrices (i.e. any two preference orderings); then, the preference ordering (theoretically existing) that has the smallest weighted sum of distances from all influencing agents' preferences is the resulting preference for the influenced agent. As the weights of the influencing agents can be positive or negative (e.g. as friends or enemies) in a real-world situation, it will partly play a role in finding the "closest" possible preference ordering from the positively influencing agents' preferences and partly play a role in finding the "farthest" possible preference ordering from the negatively influencing agents' preferences.

Most works that discuss multiple influences consider influence to be individual, which assumed that all influencing agents exert their own influences independently (from each other) and that the resulting preference/choice of the influenced agent could be a simple linearly weighted aggregation of the preferences/choices of all influencing agents. Some works discussed the influence of a coalition of multiple agents on an agent. As some agents hold the same or similar beliefs, opinions or choices (such as in an "opinion alliance"), an extra influencing effect in addition to that of the separate individual influences should be considered. However, another important influencing effect, which can be named *structural influence*, has been largely ignored. The structure here mainly refers to the influencing relationships among agents (which can be represented as links between nodes in a social network). In fact, previous work considers the structure (links) among agents just as the path or channel of influence, but relatively ignores the fact that the structure itself can also exert an additional influence as the source or origin of influence, that is, an agent may be affected not only by multiple agents' individual influences, but also by the influencing relationships among these influencing agents. However, it is not easy to address the influence of structures on an agent: as the influencing subject and the influenced object are disparate; the former is the inter-relationships between agents, while the latter is the preference/choice of a single agent. In this book, we provide a framework of the *three levels of influence* and its mathematical models to address *individual, coalitional and structural influence* and their mixed effects in the context of group decision-making.

We further consider settings of combinatorial and collective decision-making where a set of agents make choices on a set of issues in sequence based on their preferences over the set of alternatives for each issue. While agents have their initial preferences on issues, they can interact with each other, have full motivations to persuade and influence others, and be influenced by others accordingly, consequently changing their preferences/choices on these issues in the process of decision-making. The influence among multiple agents making decisions on a single issue and the dependency (influence) among multiple issues decided by a single agent have been fully discussed in previous work (such as in social influence models and CP-nets, respectively), while the influence from multiple sources across both multiple agents and multiple issues in the context of combinatorial and collective decision-making has been relatively ignored, particularly with varied strengths of influence and opposite polarities of influence. In this book, we provide three framework models to address the influence across multiple agents and multiple issues: *multiple weighted influences*, *one dominant influence*, and *two opposite influences*.

This book integrates, extends, and systematizes three papers (How to Address Multiple Sources of Influence in Group Decision-making?: From a Nonordering to an Ordering Approach; Individual, Coalitional and Structural Influence in Group Decision-making; Influence Across Agents and Issues in Combinatorial and Collective Decision-making) we have published in Springer computer science proceedings LNCS and LNBIP into a monograph.

Beijing, China   Hang Luo
May 2025

Acknowledgements This study is supported by a National Natural Science Foundation of China grant (72374010).

Contents

Introduction

The influence among behaviors and preferences in a multi-agent (such as group decision-making[1]) system[2] is quite common[3] and has been noted and discussed by scholars from various disciplines, including computer science, artificial intelligence (particularly

---

[1] Decision-making, as the core activity of social processes or procedures, touches almost every aspect of human societies, in economic fields such as how to allocate resources, what categories and quantities of goods to produce, purchase and exchange, in political fields such as who to choose as president or governor among the candidates. Especially in the discipline of management, decision has been given the most important role, considered by Herbert Simon to be the core activity of all management procedures. For example, making strategic plans for an organization, making hiring or layoff plans, and designing a salary or compensation system are all essentially making decisions. In particular, due to the social nature of human beings, group decision-making (making decisions together by more than one person) is widely required by the system of democracy, the modern enterprise institution, the need to share risk, the bond of affection, etc.

[2] Multi-agent system is an important branch of distributed artificial intelligence.

[3] People may observe the made decisions of others or communicate with others about planned decisions, they may deliberate, they may exchange arguments before making decisions. One reason for the prevalence of influence in group decision-making is the push of tools and technologies that enable massive online communication. Such an environment typically allows agents to access new information or exchange information before taking actions. In such an environment, the fact that an agent's decision may be influenced by others is the norm rather than the exception.

© The Author(s), under exclusive license to Springer Nature Switzerland AG 2026
H. Luo, *Influence Models in Group Decision-Making*, Synthesis Lectures on
Computer Science, https://doi.org/10.1007/978-3-032-01352-1_1

multi-agent system), economics, and decision theory, etc. [11, 13, 14, 17–25, 34, 37, 42, 43, 45, 48, 49, 55, 58]. In the context of a group decision (such as voting,[4] election[5]), an agent (a general term[6] that can represent a person or an artificial intelligence[7] in nature) always has the motivation[8] to influence[9] other agents regarding their choices or preferences so that they support his or her own preferred alternative (candidate), thus increasing the possibility that his or her preferred alternative will become the result of the collective choice, or so that they oppose his or her own hated alternative (candidate), thus decreasing the possibility that his or her hated alternative will become the result of the collective choice. Further, influences in real-world settings are diversified in both polarity and strength, such as a positive influence[10] from a friend (ally) versus a negative influence[11] from an enemy (opponent) [43] and a strong influence from an intimate friend (or family member, relative) versus a weak influence from an ordinary (casual) friend. Actually, the weight (including both strength[12] and polarity[13]) of influence can be affected by many factors, for example: the comparison between the actual or perceived[14] power (or authority) of the influencer and that of the influenced one[15] [42]; or how is the relationship (good or bad, close or distant)

---

[4] The most typical and widely used group decision method is voting, which has a long history dating back to ancient Greek city-states like Athens. Besides, bargaining and negotiation, arbitration and mediation also belong to group decision method. There is a famous bargaining model and solution proposed by John Nash. In a sense, game, particularly cooperative game, can also be understood as group decision-making.

[5] Voting in political fields is usually for elections.

[6] That can represent a decision-maker, a voter, a negotiator, or a game player, etc. in function.

[7] Including intelligent robotics [59].

[8] People have various reasons and motivations for influencing others, such as economic interests, political claims, and religious beliefs. For example, you may try to persuade others to believe what you believe (or what you want them to believe), to support what you support, or oppose what you oppose, in order to advance your own interests or utilities.

[9] Say in the form of convincing, bribing, etc.

[10] For instance, while your family members and close friends feel happy or sad, you may also feel happy or sad, which means that your feelings are positively influenced by your family and friends. For another instance, while your family members and close friends like or dislike something, you may also like or dislike the same thing, which means that your preferences are positively influenced by your family and friends.

[11] For instance, while your enemies and opponents feel sad, you would feel happy, or vice versa, which means that your feelings are negatively influenced by your enemies and opponents. For another instance, while your enemies and opponents like something, you would dislike it, or vice versa, which means that your preferences are negatively influenced by your enemies and opponents.

[12] i.e. the absolute value of the weight of influence.

[13] i.e. the sign of the weight of influence.

[14] Usually, from the perspective of the influenced one.

[15] Usually, the larger the power of the influencer, the larger the strength of the influence, and vice versa; the smaller the power of the influenced one, the larger the strength of the influence, and vice versa.

between the influencer and the influenced one[16] [45]; or how much the influenced one trusts (or believes) the influencer, not only due to the relationship with the influencer,[17] but also due to the expertise and knowledge of the influencer, considering that the higher the trust, the greater the weight of the influence [25, 45]; or how much common values the influencer and the influenced one share,[18] where the weight of the influence between the two agents can be measured by aggregating the various points of agreement/disagreement between the two agents' ideological spectrums,[19] considering that the more points they agree on, the greater the weight of the influence. Moreover, the influence of reality faced by an agent is usually not from only one single agent (at a time), but simultaneously from more than one agent (at the same time).

Previous work discussed at length influence among agents (typically as decision-makers), such as the influence of one agent at a time on another agent, or the simultaneous influence of more than one agent, but mostly regarding cardinal utility, belief, opinion, or choice [13, 14, 17–21, 23–25, 34, 37, 42, 43, 45, 48, 49, 58],[20] while less regarding ordinal preference. Actually, in the above mentioned work, agents' preference orderings (rankings) over alternatives are not specifically presented and discussed; thus, influences among agents can only work in a nonordering form from the influencing agents' (cardinal) utilities, beliefs, opinions or choices[21] to the influenced agent's (cardinal) utility, belief, opinion or choice, but not from the influencing agents' (ordinal) preferences[22] over all alternatives to the influenced agent's (ordinal) preference. Therefore, most influence models only support the

---

[16] Usually, the better the relationship, the more likely it is that the influencer will have a positive influence on the influenced one, and the worse the relationship, the more likely it is that the influencer will have a negative influence on the influenced one; the closer the relationship, the more likely it is that the influencer will have a stronger influence on the influenced one, and the more distant the relationship, the more likely it is that the influencer will have a weaker influence on the influenced one.

[17] i.e. the goodness and closeness of the relationship between the influencer and the influenced one.

[18] This factor can be understood as the similarity of preferences between the influencer and the influenced one, considering that the more similar their preferences, the greater the weight of the influence, since people usually tend to listen to the opinions that are the same or similar to their own and to resist the opinions that are different from their own [45], in order to maintain self-confidence or prevent self-skepticism and anxiety. This factor can also be understood as the compatibility of objectives (or purposes, or interests) between the influencer and the influenced one, considering that the more contradictory their objectives, the more likely it is that the influenced one will receive a negative influence from the influencer [45].

[19] Suppose there are m essential points of consideration, and the two agents agree on half of them; then, the weight of the influence between them will be $\frac{m}{2}$.

[20] Much of the work, such as [20, 21, 23, 24] assumes that choices made or opinions possessed by agents are binary (e.g. yes or no), which can be denoted by 1 and 0. [42] used three voting probabilities (for yes, abstention and no), which sum to 100%, to represent voting preference.

[21] If it is a nonranked choice about the most preferred alternative and there are m alternatives, then it is a 1-of-m choice, i.e. there are m possibilities of choice.

[22] i.e. preference orderings. If it is a ranked choice and there are m alternatives, then there will be $m!$ possibilities of ordering over all the alternatives. Usually, the computational complexity of ranked voting is much higher than that of nonranked voting for the same number of alternatives.

influence study in the context of a nonranked[23] group decision (such as using plurality rule or majority rule) but not a ranked[24] group decision (such as using Borda count or Condorcet method). However, it is widely recognized that nonranked group decision (especially voting) methods have considerable drawbacks compared with ranked group decision methods, because most of the preference information will be ignored[25] and controversial winners are easy to produce. In addition, previous work has usually considered the variation in the strength of different influences but rarely considered the variation in the polarity of different influences [11, 13, 14, 17–21, 23–25, 49, 55, 58],[26] which is similar to assuming that all influences are positive. In fact, however, both positive and negative influences are common,[27] and together constitute the real-world influence. Therefore, the first task of this book is to address influences of multiple agents simultaneously in an ordering (more specifically, ordinal preference) approach, especially with varied strengths (stronger or weaker) and opposite polarities (positive or negative), in order to obtain the resulting preference of the influenced agent.

Moreover, previous work extensively discussed influences of multiple agents, but nearly all assumed that the influencing agents exert their own influences independently from each other. Some previous work discussed the influence of a coalition of agents [20–24] possessing the same belief, opinion or choice on an agent. Considering this, the second task of this book is to display a sophisticated model of influence, *three levels of influence*, where not only influencing agents as individuals and their coalitions,[28] but also the structures among the influencing agents are considered in the processing of influence. The structures here in

---

[23] Also called nonordering.

[24] Also called ordering-based (ranking-based).

[25] In a nonranked group decision (especially voting) method, usually only the information about the top/most preferred alternatives (candidates) of all agents will be collected, while all other preference information, particularly the information about the bottom/least preferred alternatives (candidates), which is also very important, (in a sense) even more or at least equally important compared to the top preference information, is ignored. There are times when we do not care who wins, as long as that guy does not win. In some sense, the preference information closer to the top and bottom is more important. While under a ranked group decision method, all preference information has to be dealt in order to determine the winner or the result. Thus, controversial winners or unfair results are much easier to produce with nonranked group decision methods than with ranked methods. See the series of classic examples [38] proposed by Charles Dodgson; he is perhaps better known by his pen name Lewis Carroll, the author of Alice's Adventures in Wonderland.

[26] Reference [22] distinguished between positive and negative influences, but considered the influence of a coalition on an agent (from the beginning) rather than that of multiple individuals; [42, 43] considered the polarity (positive or negative) of individual influences, but in a cardinal form (e.g. voting probability) rather than an ordinary form (e.g. preference ordering).

[27] Even in a group with members close to each other such as a family, influences from different members can also be distinguished by strength and even polarity, since not all relationships in a family are harmonious at all times (such as the relationship between the wife and her mother-in-law) [45] not to mention the competitive relationships in a company, an international organization, etc.

[28] i.e. the coalitions formed by the influencing agents possessing the same or similar beliefs, opinions or making the same choice.

the context of a group decision with mutual influence indicate the influencing relationships among agents, which can be represented by links in a social network. In fact, the structures should not be perceived as just paths or channels of influence, but they themselves can exert some specific influencing effects as sources or origins of influence, which can affect the result of influence significantly. However, this *structural influence* has been nearly ignored in previous work. Formally representing[29] the *three levels of influence* and accurately computing the result of *structural influence* is the essential work in this book. It should be noted that how to address the influencing effect from structures to an agent[30] is not a straightforward question,[31] as the influencing subject and the influenced object are disparate. The former is the interpersonal relationships among agents (usually expressed by links in a social network[32]), but the latter is the preference (such as utility, belief or opinion, which is usually expressed by cardinal values,[33] which can also be an ordering over a set of alternatives) or choice (usually expressed by an alternative out of a set of alternatives) of a single agent. It is relatively simple to address the influence of multiple agents' preferences/choices on another agent's preference/choice, but how to address the influence of structures among agents on an individual agent's preference/choice to bridge the gap and achieve transformation[34] between these two disparate things[35] is a complicated but critical question.

Furthermore, we need to consider settings of combinatorial and collective (i.e. multiissue and multi-agent) decision-making, where a set of agents make choices pertaining to a set of issues in sequence based on their preferences over a set of alternatives for each issue. While agents have their initial preferences on a series of issues, they may interact with each other, be fully motivated to influence others, and, accordingly, be influenced by others,[36] consequently changing their preferences and ultimately their choices on these issues in the process of decision-making.[37] The influence on and updating of preferences or choices is usually achieved via the exchange and diffusion of information among agents and across issues.[38] The information that agents exchange or access can be from other

---

[29] Including graphical and mathematical expressions.

[30] More specifically, the agent's choice or preference.

[31] In particular, how to formally represent this special influence and compute the result of this special influence in mathematics.

[32] From the perspective of graph theory, a network is a graph.

[33] e.g. each alternative is assigned a utility value.

[34] Particularly in mathematics.

[35] From the perspective of graph theory, structures are composed of links, but a choice or preference is an attribute of a node (agent).

[36] Because it is a mutual influence.

[37] The mutual influence continues until hopefully the agents reach a stable state and declare their final preference or choice; at that point, a voting rule can be used to aggregate the agents' preferences or choices and generate the collective preference or choice [48].

[38] i.e. among different agents making decisions on different issues.

agents' observable (decision-making) behaviors[39] or from others agents' declared or shared preferences (underlying their choices).[40] In fact, besides the influence among multiple agents making decisions, the dependency (which can be understood as a special kind of influence) among multiple issues for decision making is also very common in reality, and has first drawn the attention of computer scientists [3, 4, 26, 56, 62, 63]. When an agent is making decisions on a series of issues, it is natural for him or her to refer to his or her own choices in the past regarding the same or similar issues. It means that an agent's preferences/choices on later issues are usually dependent on (or affected by) his or her own choices on prior issues. Similar to influence among agents, there are both positive and negative dependencies among issues: An agent will refer positively to (usually, make the same choice as) his or her satisfactory decisions in the past but will refer negatively to (usually, make the opposite choice to) his or her regrettable decisions from the past. Similar to influence among agents, there are both strong and weak dependencies among issues: An agent will refer more to his or her important[41] decisions and less to his or her ordinary[42] decisions in the past. It is also common that an agent's choice/preference on an issue will be influenced by another agent's choice/preference on another issue, or even by different agents' choices/preferences on different issues at the same time. Therefore, the third task of this book is to discuss the influence across both multiple agents and multiple issues in the settings of combinatorial and collective decision-making.

In fact, influence in present-day society has become much more intensive,[43] particularly through large-scale online communication via the Internet, which transcends limitations of space, time and environment [41]. More specifically, with the advancement of wireless network technology and mobile communication devices,[44] especially with the help of online social platforms, such as WeChat®, X®, and Facebook®, interaction and communication among people (particularly those in remote locations) have become much more convenient and frequent than before[45] [41]; interaction and communication are usually the foundations of influence. If you want,[46] you can instantly communicate with friends or clients overseas,

---

[39] Reference [48] assumed that the information agents exchange is the mere observation of others' choices: agents may revise their choices on the basis of the observed choices of others.

[40] In a word, agents may update their preferences or choices based on observed, declared or shared information about preferences or choices of others.

[41] Or strategic.

[42] Or daily.

[43] As a matter of common sense, if there is a society of more than one person, then there will always be influences among these persons, just on different aspects or to different extents.

[44] e.g. mobile phone, tablet computer.

[45] In an agricultural society, or before the invention of telecommunication and network technology, people could only frequently interact with their neighbors (in the sense of spatial distance). In today's society, interaction and communication among people are not limited to space and environment anymore.

[46] On most cases, determining whether two persons interact and communicate with each other is the willingness to interact rather than spatial distance, technology and devices. Nowadays, most people

no matter how big or small.[47] Staying in touch[48] with friends who are far away is no longer a problem[49] in present society. Thus, current studies of decision-making should typically involve a large number of agents interacting with each other and making decisions on a series of issues[50] as opposed to restricted cases consisting of a few agents making decisions on a single issue, or a few issues decided by a single agent. In this context, the interaction and influence among different agents on different issues should be fully investigated, which makes the study of psychology and behavior of decision-makers, the mechanisms and dimensions of influence, and the aggregation for collective preference/choice in group decision-making more complicated.

There is a general meaning for the study of influence in group decision-making, not only with regard to theory of computer science and artificial intelligence (particularly multi-agent system), economics and management (particularly social choice,[51] and

---

have mobile phones and use online social platforms. If you want to communicate, you can chat with your friends, family, or clients overseas and anywhere; if you do not want to communicate, you would even not say hello to a next-door neighbor, particularly in megacities such as Beijing and New York.

[47] i.e. whether it is a big thing or a small thing.

[48] Even all the time.

[49] In the sense of space, time and environment.

[50] Which is usually full of influences and dependencies.

[51] It is specially meaningful for computational social choice [9, 10, 12, 57] which is an interdisciplinary research field between computer science (particularly computational complexity), artificial intelligence (particularly multi-agent system) and social choice theory. Specifically, it is the study of social choice, particularly voting and fair resource allocation, from the perspective of computer science. Computational social choice aims at studying the computational aspects of collective decision-making, including complexity analysis and algorithm design, more specifically, how to measure and evaluate the communication complexity of a fair (resource) division algorithm, the computational complexity of determining the winner for a given voting rule and that of manipulating the winner for a given preference distribution, and how to design rules and use computational complexity to defend against strategic voting and manipulation (usually, strategic voting is the means of manipulation and manipulation is the objective of strategic voting). Computational social choice combines both technological (computational) complexity and social complexity (particularly of human psychology and behavior in decision-making).

group decision[52]), etc., but also as it pertains to application[53] in the committee in all kinds of organizations, joint-stock company voting, domestic political elections,[54] and international organization decision-making,[55] etc.

---

[52] Social choice theory and group decision theory overlap considerably. Economists and political scientists may prefer the term "social choice", while management scientists, particularly those specializing in decision theory, may prefer the term "group decision".

[53] The models of influence can also be combined with agent-based modeling and simulation (ABMS) for test and experiments.

[54] For presidents, governors, senates, representatives, etc.

[55] Almost all international organizations adopt group decision-making systems, whether they are international political (or comprehensive) organizations such as the UN General Assembly, the UN Security Council [43] and the European Union [42], or international economic (or financial) organizations such as the World Bank and the International Monetary Fund (IMF) as the Bretton Woods institutions and the Asian Infrastructure Investment Bank [47] and the New Development Bank [46] as emerging multilateral financial institutions.

Background

2

2.1   Social Choice Theory

Social choice refers to combining (aggregating) individual choices, preferences or welfare to reach a collective choice, preference or social welfare.[1] Social choice functions are those functions that take individual choices/preferences as input and give a collective choice/preference as output. Social choice methods[2] include any method that can help determine the social choice/preference based on individual choices/preferences. Social choice methods mainly include group decision (most typically, voting), negotiation, etc. A widely used classification of social choice functions or methods considers ranked and nonranked methods. In nonranked methods, each individual can only present his or her most preferred alternative (candidate),[3] whereas ranked methods allow a full ordering among all alternatives (candidates) to be displayed by each individual.

In this section, we will review typical nonranked and ranked social choice functions and show why it is widely recognized that ranked social choice methods tend to produce a fairer result compared to nonranked social choice methods, and thus are more reliable.

---

[1] Social choice theory refers to a theoretical framework for the analysis of the aggregation process.

[2] Social choice functions are the mathematical expressions of social choice methods.

[3] i.e. each individual chooses one alternative out of the set of alternatives. In this book, we focus on voting methods to produce or select one winner rather than multiple winners, since we will extend social choice methods or functions to *social influence functions* where there should be only one winner of influence.

© The Author(s), under exclusive license to Springer Nature Switzerland AG 2026
H. Luo, *Influence Models in Group Decision-Making*, Synthesis Lectures on
Computer Science, https://doi.org/10.1007/978-3-032-01352-1_2

2.1.1 Nonranked Social Choice Functions

Plurality and majority are the two most typical and widely used nonranked social choice methods.

Plurality is the most simple and natural solution to choose a winner: each agent casts one vote for (only) one alternative (candidate), and the alternative with the highest number of votes wins [30].

Majority is relatively stricter, requiring an alternative (candidate) to receive more than half of the votes, not just the highest number of votes, to be the winner [30]. If no alternative can obtain more than half of the votes, the voting process cannot be finished (in just one round). There are some remedies for this situation, such as the two-round system (also called the second ballot). It is possible that an alternative can win by plurality rule but not by majority rule.

We use a classic example [38] proposed by Charles Dodgson to illustrate how the two methods work; he assumed a group decision-making (voting) system composed of a set of voters as $N = \{1, 2, ..., 11\}$ and a set of alternatives (candidates) as $M = \{A, B, C, D\}$, i.e. there are 11 voters choosing from 4 alternatives, with the distribution of the preference ordering of all voters as shown in Fig. 2.1. If plurality rule is adopted and each voter votes for his or her most preferred alternative (candidate), then A receives 3 votes, B receives 4 votes, C receives 3 votes, and D receives 1 vote, thus B, who receives the highest number of votes, is the winner. But is this result fair? In fact, we can easily see that, except for 4 voters (No. 4, 5, 6, and 7) who put B in the first preferred position in their preference orderings, all other 7 voters (No. 1, 2, 3, 8, 9, 10, and 11) put B in the least preferred position in their preference orderings (i.e. rate B the worst). Therefore, the result of the collective choice is

No. of Voters		1	2	3	4	5	6	7	8	9	10	11
Preference Ordering	1st	A	A	A	B	B	B	B	C	C	C	D
	2nd	C	C	C	A	A	A	A	A	A	A	A
	3rd	D	D	D	C	C	C	C	D	D	D	C
	4th	B	B	B	D	D	D	D	B	B	B	B

Fig. 2.1 An example proposed by Charles Dodgson why nonranked social choice (voting) is unfair or unreliable. *Notes* Each column represents the preference ordering among all alternatives of a voter. For instance, the first column with colors means that the voter (No. 1) puts A in the first place (i.e. prefers A the most), puts C in the second place, puts D in the second last place, and puts B in the last place (i.e. prefers B the least). The grids representing the first place are colored in green, the grids representing the second place are colored in light green; the grids representing the last place are colored in red, the grids representing the second last place are colored in light red

No. of Voters	1	2	3	4	5	6	7	8	9	10	11
1st	B	B	B	B	B	B	A	A	A	A	A
2nd	A	A	A	A	A	A	C	C	C	D	D
3rd	C	C	C	D	D	D	D	D	D	C	C
4th	D	D	D	C	C	C	B	B	B	B	B

Fig. 2.2 Another example proposed by Charles Dodgson why nonranked social choice (voting) is unfair or unreliable

quite controversial, and would be considered unfair by the majority of voters.[4] In fact, A may be a fairer choice as the winner, since A is ranked in the top 2 by all voters.

Some people may say that the requirement of plurality rule is too loose, and once majority rule is adopted, requiring the winner to get more than half of the votes, there will be no such unfair result. Let us consider another example [38] proposed by Charles Dodgson with the distribution of the preference ordering of all voters as shown in Fig. 2.2, in which there are still 11 voters choosing from 4 alternatives (candidates). According to majority rule, A receives 5 votes, B receives 6 votes, thus B, who receives more than half of the votes, is the winner. But this result may still be unfair, since nearly half of the voters (No. 7, 8, 9, 10, and 11) rank B at last (i.e. rate B the worst). The result of the collective choice is still quite controversial, possibly strongly disagreed by almost half of the voters.[5] In fact, A may be a fairer choice as the winner, since A is ranked in the top 2 by all voters.

2.1.2 Ranked Social Choice Functions

Borda count and Condorcet method are the two most classic and typical ranked social choice methods.

Borda count allows each agent to rank all alternatives (candidates) on his or her ballot.[6] The alternatives each obtain a number of scores based on their ranks on all of the agents' ballots according to a score allocation scheme, where the scores decrease with respect to the rank [30]. A typical score allocation scheme is as follows: if there are m alternatives,

---

[4] It is okay if the majority is just unsatisfied with the result but can still accept it, it is troublesome if they do not recognize the result, then it is possible for the group or organization to divide and disintegrate.

[5] A similar example in real-world politics is the 2016 and 2024 US presidential elections, which were won by Donald Trump. Half of Americans like Trump so much, while the other half of Americans hate him so much.

[6] i.e. it allows each agent to provide a full ordering among all alternatives.

the rth ranked alternative is allocated a score of $m - r + 1$, i.e. scores of $m, m - 1, ..., 1$ correspond to the 1st, 2nd, ..., and mth ranked alternatives, respectively[7] [30]. The scores that each alternative receives from all the agents' ballots are summed, and the alternative with the highest summed score is declared the winner.

The score counted in Borda count is usually called the Borda score. We can define the Borda score for each alternative as $\mathbf{B}_o$ ($o \in \mathbb{M}$ represents an alternative in the set of alternatives). According to the preference distribution example shown in Fig. 2.2 and adopting the score allocation scheme of 4, 3, 2, and 1 to be assigned to the 1st, 2nd, 3rd, and 4th ranked alternatives, respectively, there will be: $\mathbf{B}_A = 4 \times 5 + 3 \times 6 + 2 \times 0 + 1 \times 0 = 38$, as there are 5 voters (No. 7, 8, 9, 10, and 11) ranking A in the first place, 6 voters (No. 1, 2, 3, 4, 5, and 6) ranking A in the second place, and none ranking A in the last two places; $\mathbf{B}_B = 4 \times 6 + 3 \times 0 + 2 \times 0 + 1 \times 5 = 29$, as there are 6 voters (No. 1, 2, 3, 4, 5, and 6) ranking B in the first place, none ranking B in the second and third places, and 5 voters (No. 7, 8, 9, 10, and 11) ranking B in the last place; $\mathbf{B}_C = 4 \times 0 + 3 \times 3 + 2 \times 5 + 1 \times 3 = 22$, as there are none of the voters ranking C in the first place, 3 voters (No. 7, 8, and 9) ranking C in the second place, 5 voters (No. 1, 2, 3, and 10, 11) ranking C in the third place, and 3 voters (No. 4, 5, and 6) ranking C in the last place; $\mathbf{B}_D = 4 \times 0 + 3 \times 2 + 2 \times 6 + 1 \times 3 = 21$, as there are none of the voters ranking D in the first place, 2 voters (No. 10 and 11) ranking D in the second place, 6 voters (No. 4, 5, 6, 7, 8, and 9) ranking D in the third place, and 3 voters (No. 1, 2, and 3) ranking D in the last place; since $\mathbf{B}_A > \mathbf{B}_B > \mathbf{B}_C > \mathbf{B}_D$, A is the winner instead of the controversial B (which is the winner under both plurality rule and majority rule).

Condorcet methods also allow each agent to rank all alternatives (candidates) according to his or her preference. The alternative that is pairwise preferred to all other alternatives by the majority of agents is called the Condorcet winner.[8] A Condorcet method for m alternatives in essence is to hold $C_m^2 = \frac{m(m-1)}{2}$ majority elections between all pairs of alternatives [30].

---

[7] Another typical score allocation scheme is as follows: if there are m alternatives, the rth ranked alternative is allocated a score of $m - r$, i.e. scores of $m - 1, m - 2, ..., 0$ correspond to the 1st, 2nd, ..., and mth ranked alternatives. The winners under the two schemes are the same. Both the two schemes allocate scores by constant difference. There are also some atypical Borda counts, in one of which, the rth ranked alternative is allocated a score of $\frac{1}{r}$, i.e. scores of $1, \frac{1}{2}, ..., \frac{1}{m-1}, \frac{1}{m}$ correspond to the 1st, 2nd, ..., $m - 1$th, and mth ranked alternatives. This score allocation scheme highlights the significance of first place as the score of $\frac{1}{2}$ allocated to the alternative ranked in the second place is just half of the score of 1 allocated to the alternative ranked in the first place; meanwhile, this scheme does not think that those alternatives in the bottom have substantive difference as the score of $\frac{1}{m-1}$ allocated to the alternative ranked in the second last place is nearly the same with the score of $\frac{1}{m}$ allocated to the alternative ranked in the last place. Such a design makes sense, for example, in the Olympic Games, a gold medal is indeed far more valuable than a silver medal, while the third-last and second-last are nearly indifferent.

[8] Condorcet method is widely recognized as the most strict and rigorous standard. We can image a candidate who can beat any other candidate by one-on-one comparison, if he or she is not the winner, then who is qualified to be the winner?

To do pairwise comparison, we need to define the individual preference between two alternatives as $\succ_i$ ($i \in \mathbb{N}$ represents an agent in the set of agents, $\succ$ is the symbol of "superior"[9]), define $N_{o \succ_i o'}$ as the number of agents preferring alternative o to alternative o', and then define the collective preference between two alternatives aggregated from individual preferences as $\succ_G$ (G means general or group). If $N_{o \succ_i o'} > N_{o' \succ_i o}$, then the group prefers alternative o to alternative o'; if $N_{o' \succ_i o} > N_{o \succ_i o'}$, then the group prefers alternative o' to alternative o; if $N_{o \succ_i o'} = N_{o' \succ_i o}$, then the group considers that the two alternatives o and o' are indifferent. In a pairwise comparison between two alternatives, only one alternative can reach a majority. According to the preference distribution example shown in Fig. 2.1, there will be: $N_{A \succ_i B} = 3 + 4 = 7 > \frac{11}{2} \Rightarrow A \succ_G B$, as there are 7 voters (No. 1–3 and No. 8–11) deeming A is better than B, reaching the majority (of the 11 voters in total); $N_{A \succ_i C} = 3 + 4 + 1 = 8 > \frac{11}{2} \Rightarrow A \succ_G C$, as there are 8 voters (No. 1–3, No. 4–7, and No. 11) deeming A is better than C, reaching the majority; $N_{A \succ_i D} = 3 + 4 + 3 = 10 > \frac{11}{2} \Rightarrow A \succ_G D$, as there are 10 voters (No. 1–3, No. 4–7, and No. 8–10) deeming A is better than D, reaching the majority; thus, A is the Condorcet winner, as A can "beat" any other alternative by pairwise comparison.[10]

2.2   Social Network Theory

A social network is a social structure made up of a set of social actors[11] (represented by nodes/vertexes in a graph or agents[12] in a multi-agent system) and a set of bilateral relationships between them (represented by ties/edges/links[13] in a graph). Graph theory is the mathematical foundation of social network theory, a social network is a graph from the perspective of graph theory. The field of social networks has emerged as a very successful interdisciplinary area of research,[14] drawing the attention of scholars from various disci-

---

[9] Which should be distinguished from $>$. Another related symbol is $\sim$, which means "indifference".

[10] It should be noted that there is not always a Condorcet winner and the Condorcet paradox (also called the paradox of voting) occurs (Kenneth Arrow's impossibility theorem rigorously proved it), therefore a number of methods, such as Black's method (which is a combination of Condorcet method and Borda count, uses Condorcet method first, if there is no Condorcet winner, then adopts Borda count) [30], Dodgson's method [12], Young's method [10], are proposed to find a replacement winner [30]. Usually, the idea is to find the non-Condorcet winner that is closest to being a Condorcet winner.

[11] Which are persons traditionally.

[12] These three terms (node, vertex, and agent) can express the same thing. Different researchers from different disciplines have different preferences. In the field of artificial intelligence (particularly multi-agent system), researchers usually call nodes and vertexes agents while using social network models. A social network can be a multi-agent system in their eyes [25, 58].

[13] These three terms can also express the same thing.

[14] Social network theory refers to a theoretical framework for the analysis of social networks.

plines, including sociology[15] [15–18, 27–29, 33, 34, 61], mathematics (particularly graph theory) [31], physics[16] [1, 51–54], computer science (particularly theoretical computer science and artificial intelligence)[17] [6–8, 11, 25, 45, 48, 49, 55, 58, 60], economics [14, 19–24, 37, 40], and politics (particularly international relations)[18] [42, 43], etc. Traditionally, in a social network, social actors represent persons and the relationships between them represent interpersonal relationships. Social actors can also represent organizations made up of persons and even countries [42, 43] and nowadays, artificial intelligence.[19] (Interpersonal) ties are generally varied in strength as stronger or weaker[20] [27–29, 40] and opposite in polarity, i.e. positive or negative [1, 31, 33, 42, 43, 45].

2.2.1 Weak and Strong Ties

Granovetter [29] stated, "The 'strength' of an interpersonal tie is a linear combination of the amount of time, the emotional intensity, the intimacy (or mutual confiding), and the reciprocal services which characterize each tie". He distinguished between strong ties and weak ties. Strong ties usually connect you with close friends and family members, while weak ties usually connect you with ordinary friends and casual acquaintances. However, Granovetter [29] argued that weak ties can be more important than strong ties, as strongly linked persons (such as our family members and close friends) stay in the same social circle with us, and the information and resources they possess usually overlap considerably with what we already have, whereas weak ties connect us to people from different social circles. Therefore, weak ties, bridging persons in different social circles (as shown in Fig. 2.3), are

---

[15] Also including some psychologists. One representative promotor of social network analysis in the field of sociology is Linton Freeman, who systematically studied the concept and measure of centrality in social networks.

[16] There is a bunch of physicists working on complex networks. One representative is Mark Newman, who is famous by studying scientific collaboration networks.

[17] One representative promotor of social network analysis in the field of computer science is Ulrik Brandes, who and some other scholars introduce the concept and model of (electrical) current flow into network analysis, referring to Kirchhoff's Current Law and Kirchhoff's Voltage Law, to provide current-flow variants of betweenness centrality and closeness centrality which can address weighted networks effectively.

[18] Traditionally, social network analysis studies interpersonal relationships. In fact, social network analysis can naturally be applied to the study of international relations, where each node represents a country and each link represents the bilateral relationship between two countries.

[19] Especially those artificial intelligences with common sense [5].

[20] Or even be absent. According to Mark Granovetter [29], "absent" ties means lack of any relationship or those relationships without substantial significance, such as "nodding" relationships between people living on the same street, or the "tie" to a vendor from whom one frequently buy groceries. The fact that two people know each other by name does not necessarily qualify the existence of a weak tie. If their interaction is negligible, the tie may be absent.

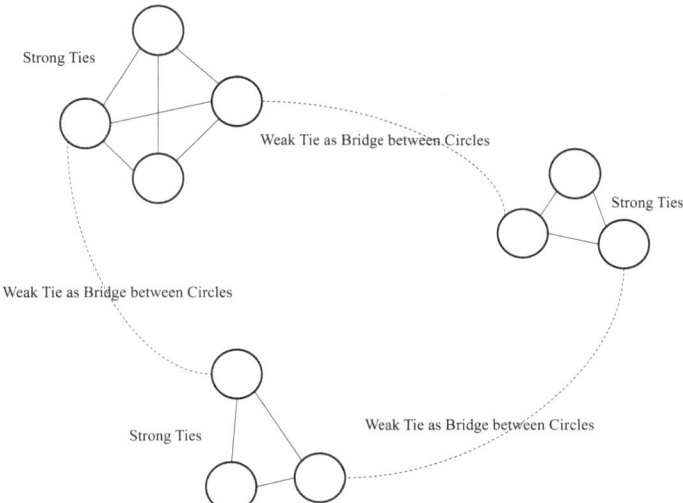

Fig. 2.3 Weak ties as bridges. *Notes* We use solid lines to represent strong ties and dashed lines to represent weak ties. "Bridge" is actually a technical term used in social network analysis to refer to a tie that, if removed, breaks a network into two disconnected sub-networks. In another word, a bridge is a type of social tie that connects two different circles in a social network. Bridges transmit new information from one circle to another. The breadth of information spread depends heavily on the number and connectedness of the bridges available to the originators of the information

responsible for the majority of new information transmission[21] and resource exchanges, and such an idea is concluded as the *strength of weak ties*[22] [29].

Krackhardt [40] proposed a contrary idea of the *strength of strong ties*, emphasizing the advantage of strong ties,[23] and stated, "People resist change and are uncomfortable with uncertainty. Strong ties constitute a base of trust that can reduce resistance and provide comfort in the face of uncertainty",[24] thus he argued that change[25] is not facilitated by weak ties but rather by strong ties. Particularly when it comes to major change, which may threaten

[21] Specifically speaking, more new information flows to individuals through weak ties rather than strong ties.

[22] In a survey he conducted of how 282 persons in the United States got their jobs, Granovetter found that a person's weak ties (their casual connections and loose acquaintances) were more helpful than his or her strong ties in securing employment.

[23] David Krackhardt [40] called strong ties "philos" and defined philos relationships as those relationships that satisfy the following three necessary and sufficient conditions: the first one is interaction, for i and j to be philos, i and j must interact with each other; the second one is affection, for i and j to be philos, i and j must feel affection for each other; the third one is time, for i and j to be philos, there must be a history of interactions that have lasted over time.

[24] Which means that strong ties are crucial during times of severe change and uncertainty.

[25] Including reform and even revolution, etc.

the status quo in terms of power and the standard routines of how decisions are made, trust is required; thus, change is the product of strong ties [40]. It is easy to understand that people usually trust close friends and family members more than ordinary friends and causal acquaintances, which means that people usually trust their strong ties more than their weak ties. Although weak ties are a powerful way to convey awareness of new things, they are weak at conveying behaviors that are in some way risky or costly to adopt [40]; however, such behaviors can be crucial to reform and innovation. In fact, Granovetter [27] also admitted that "weak ties provide people with access to information and resources beyond those available in their own social circle; but strong ties have greater motivation to be of assistance and are typically more easily available". Afterall, weak ties are usually more unreliable (undependable) and easy to break in a turbulent and risky environment. When we emphasize the value of weak ties for introducing new information from different social circles, we should also acknowledge the fragility and distrust associated with weak ties. As an ancient Chinese proverb states, "the poor in the downtown area go unnoticed, the rich in the mountains have distant relatives".

2.2.2 Positive and Negative Ties

Interpersonal ties can be positive or negative. There are positive ties with friends and family members, etc.; however, it is also possible to face negative ties in real-world settings with enemies and opponents.[26] The most classical work addressing positive and negative ties is *structural balance theory* proposed by Heider[27] [33], which discussed the balance or unbalance of a triangular relationship (among three persons) composed of three bilateral relationships (ties) that can be either positive or negative (respectively representing amity or enmity relationships). The product of the signs (positive or negative) of the bilateral relationships (ties) of a triangular relationship determines whether it is balanced or unbalanced such that if the product is negative,[28] then the triangular relationship is unbalanced; if the product is positive,[29] then the triangular relationship is balanced [33], which is based on cognitive transitivity[30] or, more simply, for being balanced, the ancient proverbs, "My friend's friend is my friend", "My enemy's enemy is my friend", "My enemy's friend is my enemy", and "My friend's enemy is my enemy", should be obeyed rather than violated.

Further, Harary[31] [31] discussed networks composed of more than three nodes connected by positive/negative ties from a mathematical perspective, conducted a theoretical study of abstract representations of networks, called networks with positive and negative ties *signed*

---

[26] Or any person with a negative view of each other.

[27] The well-known attribution theory was also proposed by Fritz Heider, which explains how people attempt to explain the causes of certain behaviors and events.

[28] Or more specifically, if the number of negative ties is odd (1 or 3).

[29] Or more specifically, if the number of negative ties is even (0 or 2).

[30] Or consistency.

[31] Frank Harary is one of the founders of modern graph theory.

graphs, and determined the conditions for a signed graph to be balanced: only if each of the cycles within the signed graph is balanced. A cycle refers to a closed walk through at least three nodes via links (ties), in which all links are distinct and all nodes except the first and last nodes are distinct [61]. A cycle is balanced if the product of the signs of all links (ties) in this cycle is positive. To simplify the determination of balance of a signed graph and reduce the computational complexity, [60] directly examined whether each triangular relationship (triangle) in a signed graph is balanced: only if every triangle in this signed graph is balanced,[32] then the whole graph is balanced; if at least one triangle in this signed graph is unbalanced,[33] then the whole graph is unbalanced. Harary [31] also proved an easier approach to determine whether a signed graph is balanced: a signed graph is balanced only if its nodes can be separated into two mutually exclusive subsets[34] such that each positive tie joins every two nodes within the same subset and each negative tie joins every two nodes from different subsets;[35] one of the two subsets can be empty (then all nodes are joined by positive ties).[36]

2.2.3   Inward and Outward Ties

We should distinguish between undirected and directed networks. In undirected networks, there is no direction for links (ties), whereas in directed networks, links (ties) are directed and the directional information should not be ignored. For instance, marriage is undirected because we should not say that the wife follows the husband or the husband follows the wife. However, love is directed, because it happens that you love someone, but he or she does not love you, which means that there is a directed link (orienting) from you to him or her, but no directed link (orienting) from him or her to you, and that is what usually makes us sad. When we consider a directed network, we should distinguish between inward and outward links; for any two nodes i and j, there may exist a link (directed) from i to j, or a link from j to i, or both links. Furthermore, the link from i to j and the link from j to i are usually asymmetric rather than symmetric. For example, even if two people love each other, the degree of their love can be quite different.

Influence networks should be directed networks, because influence has direction, we need to distinguish between the subject of influence (influencer) and the object of influence (influ-

---

[32] i.e. only if the product of the signs of the bilateral relationships in each triangular relationship is positive.

[33] i.e. if the product of the signs of the bilateral relationships of at least one triangular relationship is negative.

[34] Which means a social system is split into two cliques.

[35] The most classic example in history is the formation of two opposing camps, the Central Powers and the Allied powers (Allies), before World War I, which has been discussed by some physicists [1] from the perspective of a signed graph.

[36] Which is a special case of two mutually exclusive subsets that may occur in very small networks and is not realistic in international relations.

enced one), briefly, who influenced whom? Such a directed network is a typical multigraph because there can exist more than one link between a pair of nodes.

2.3   Social Influence Model

Social influence models capture influence or influencing relationships among agents.[37] Influence is the process or result when the behaviors or preferences[38] of agents are affected (changed) by others. Social influence may take many forms and can be seen in conformity, peer pressure, leadership, persuasion [14], etc.

In the fields of computer science and artificial intelligence (particularly multi-agent system) [11, 25, 45, 48, 49, 55, 58], economics [14, 19–24, 37], sociology [17, 18, 34], and politics [42, 43], etc., the process and result of influence has been intensively studied, especially in the context of social networks, using ties (links) between nodes (agents) to represent the influencing relationships between them. Indeed, influence is usually asymmetric [42]: there is no reason that the way[39] agent i influences agent j should be exactly the same as the way agent j influences agent i, therefore, the direction of influence should be considered. Furthermore, influence is usually mutual and iterative[40] [24]: it may well be the case that agent i influences agent j, which then influences agent k, which in turn influences agent i.

Many of these influence models assume a unidimensional value as utility or belief for each agent, which will be affected by his or her "neighbors",[41] depending on the social structure. In this line, the model of "reaching a consensus" [13] and the theory of "decisional power" [34] may be foundational. Systematically, Jackson [37] described a social network scenario where every agent's belief is influenced (in the form of learning) by the agent's own and all other agents' beliefs[42] according to a weight allocation, where the weights (of influence) may be bigger or smaller to represent stronger or weaker influences among agents. In fact, the influence among agents in a (social) network can be represented by a matrix consisting of entries that each represents the weight of influence between two agents.[43] Indeed, if we have a network (graph), we can fill in a matrix; if we have such a matrix, we can draw a network (graph).[44]

---

[37] Specifically, among agents' behaviors or preferences.

[38] Such as utilities, beliefs, or opinions.

[39] Including the strength (stronger or weaker) of influence and the polarity (positive or negative) of influence, etc.

[40] Reference [24] analyzed the group decision-making process in which the mutual influence does not stop after one step but iterates. For iterative influence, a central question has been whether a convergence or consensus state can be reached at the end.

[41] Neighbors of an agent in a social network mean other linked agents.

[42] Only if their weights of influence are not 0.

[43] The conditions on the matrix that represents influences allow to determine whether or not consensus will be reached in the end.

[44] Which means that a network and a matrix can be mathematically equivalent and transformed to each other.

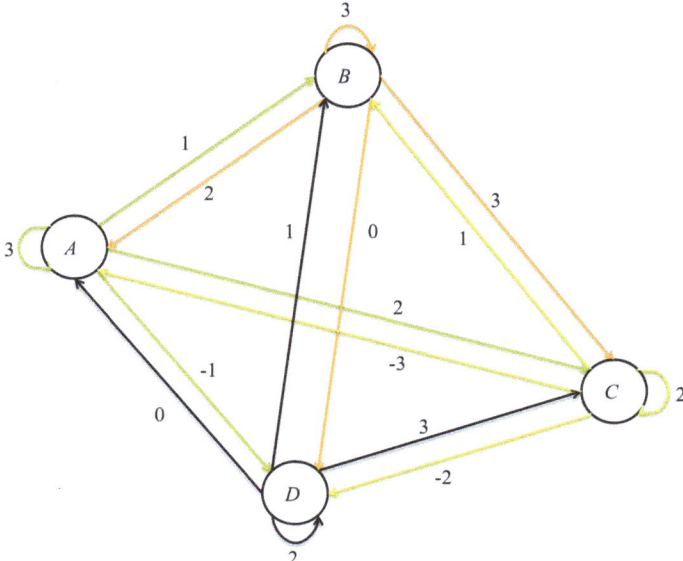

Fig. 2.4 One example of graphical representation of social influence. *Notes* Each number besides a link (directed line) indicates the weight of the influence represented by the link. The deep yellow lines represent the influences of agent *A* on other agents and himself or herself, the orange lines represent the influences of agent *B* on other agents and himself or herself, the yellow lines represent the influences of agent *C* on other agents and himself or herself, and the black lines represent the influences of agent *D* on other agents and himself or herself

Example 2.1 (Graphical and Matrix Representations of Social Influence) Assume there are four agents *A*, *B*, *C*, *D* who influence and are influenced by each other, as shown in Fig. 2.4. As mentioned, social influence among agents in the context of a social network can be represented by a matrix whose entries are the weights of influence between any two agents. The matrix representing the social influence in Fig. 2.4 is provided in Table 2.1. We define $w_{i,j}$ as the weight of influence from agent *i* to agent *j* (i.e. of agent *i* on agent *j*); the weight value indicates both the strength and polarity of the influence, $w_{i,j} > 0$ means a positive influence from agent *i* to agent *j*, $w_{i,j} < 0$ means a negative influence from agent *i* to agent *j*, and $w_{i,j} = 0$ means absent influence from agent *i* to agent *j*.

To understand the diverse influences among agents in this example, we should distinguish between influences coming from and going to certain agents: on the one hand, from the perspective of an influenced agent, say *A*, he or she thinks that agent *B* is his or her friend ($w_{B,A} = 2$, which indicates a positive influence), while agent *C* is his or her enemy ($w_{C,A} = -3$, which indicates a negative influence), and does not care about agent *D* ($w_{D,A} = 0$, which may mean that agent *D* is "nobody" in the mind of agent *A*); on the other hand, from the perspective of an influencing agent, also say *A*, he or she is counted as a friend by agents *B*

Table 2.1 One example of matrix representation of social influence

$$
\begin{array}{cccc}
 & A & B & C & D \\
A & \begin{pmatrix} 3 & 2 & -3 & 0 \\ 1 & 3 & 1 & 1 \\ 2 & 3 & 2 & 3 \\ -1 & 0 & -2 & 2 \end{pmatrix}
\end{array}
$$

Notes The horizontal axis indicates the influencing agents, and the vertical axis indicates the influenced agents

and C ($w_{A,B} = 1$ and $w_{A,C} = 2$, which both indicate a positive influence), while he or she is counted as an enemy by agent D ($w_{A,D} = -1$, which indicates a negative influence). Except for agent A, each of agents B, C, and D is both an influencing agent and an influenced agent.

In addition, the influence of an agent on himself or herself is also considered in this example ($w_{A,A} = 3$, $w_{B,B} = 3$, $w_{C,C} = 2$, and $w_{D,D} = 2$), since our current preferences or choices are usually dependent on or influenced by our own past preferences or choices. In some sense, who I am today continues from who I was yesterday.

Furthermore, it should be noted that the weights of influence between two agents may not be symmetric, such as $w_{A,C} = 2$ but $w_{C,A} = -3$, which means that agent C perceives agent A as a friend (with a positive influence) while agent C perceives agent A as an enemy (with a negative influence), this kind of "contradiction" is very common in real-world situations. For instance, sometimes you think someone is your friend, while he or she does not think so, and considers you a bother.

The study on influence by Grabisch and Rusinowska [20–24] should also be mentioned. They discussed and compared influence functions,[45] follower functions[46] [23] and command

---

[45] Previous works [23, 48] proposed some examples of influence functions and discussed their terminal states: (1) *Fol* is an influence function between two agents, each of which always follows the inclination of the other (e.g. each agent always thinks the other's inclination or choice is better than his or her own). This influence function converges to stability when the initial inclinations are a consensus between the two agents. Otherwise, the influence iteration never stops; (2) *Gur* is an influence function where one of the agents is the guru (similar to opinion leader or charisma) and all other agents follow him or her. Given any distribution of the initial inclinations, the iteration will converge to one of the stable states according to the initial inclination of the guru. Each stable state represents a consensus; (3) *Conf3* models a community with a king, a man, a wife, and a child, following a Confucian way: the man follows the king, the wife and the child follow the man, and the king is influenced by (listens to) others (In Confucianism, the state and family have the same structure. This is why it is assumed that the man listens to king, and the wife and the child listen to the man). This influence function always converges to one of the stable states, each of which represents a consensus.

[46] Grabisch and Rusinowska [23] studied the relationship between influence functions and follower functions, delivered sufficient and necessary conditions for a function to be a follower function, and described the structure of the set of all influence functions that lead to a given follower function. A follower function assigns each coalition the set of its followers.

games[47] [20], that are embedded in social networks [21]; they assumed that agents are to make a binary choice,[48] where each agent has an inclination to say[49] either "yes" or "no", and due to the influence of other agents, the choice of an agent may be different from his or her original inclination; they defined such a transformation as influence.[50] Further, they [22] generalized the yes/no binary choice model to a multi-choice framework (choosing among more than two alternatives) where each alternative is assigned a number (similar to utility value) which can be influenced by others, thus, the influencing subject and the influenced object are essentially cardinal utilities, but not ordinal preferences (in fact, it is much more demanding to ask all agents to assign a utility value to each alternative than it is to simply rank these alternatives); they [22] also discussed both positive and negative influences in a coalitional way rather an individual way, which means that in some sense they "skipped" the influence of individuals directly to the influence of a coalition, not distinguishing the influence of a coalition as an additional influencing effect on top of the individual influences; further, they [24] generalized a yes/no model of influence in a social network with a single step of mutual influence to a dynamic model with iterating influence.

More recently, [25] proposed a model combining opinion diffusion, judgment aggregation and social network, where each agent's binary (yes/no) opinions will be affected by his or her (linked) neighbors in the network according to the trust the agent has in them;[51] however, influence was still assumed to be on binary opinions (or choices) rather than on preference orderings over multiple alternatives. References [11, 55] discussed influence and its impact on the evolution of experts' opinions in the context of a social network, where opinions are represented by preference relations[52] (each expressed by a cardinal value) between any two

---

[47] Grabisch and Rusinowska [20] studied the relationship between command games proposed by Hu and Shapley [35, 36] and an influence model, showed that the framework of influence is more general than the framework of the command games, and delivered sufficient and necessary conditions for a function to be a command function. In a command game, for each player, boss sets and approval sets need to be introduced; the boss sets are defined as the sets of individuals that the player must obey, regardless of his or her own judgment or desires; the consent of the approval set is sufficient to allow the player to act [35].

[48] Acceptance or rejection.

[49] More formally, vote.

[50] "Each player makes his [or her] decision which, due to the influence of other players, may be different from the original inclination of the player. Such a transformation of the inclinations into the decisions is represented by an influence function" [22].

[51] In general, the more trust an agent has in a neighbor, the more influence that neighbor will have on the agent.

[52] Preference relation is based on the idea of pairwise comparisons: experts' preferences are described by preference relations in which each value assigned to a preference relation between two alternatives represents the preference of one alternative over the other; whereas in the case of preference ordering, experts provide their preferences for all alternatives as an ordered vector of alternatives, from the best one to the worst one [55].

alternatives, which are still not (pure) ordinal preference over all alternatives; usually, it is much more demanding to ask all experts to give a value for each preference relation between any two alternatives (there are m^2 pairwise comparisons to be each assigned a value if there are m alternatives,[53] which indicates a $m \times m$ matrix to fill in) than it is to simply give an ordering among alternatives (equivalent to filling in a row vector of length m with m alternatives without the requirement to assign a value for each pairwise comparison); furthermore, although they all discussed individual influences, negative influences of neighbors were not considered, rather, it was assumed that all neighbors' opinions have positive influences.

2.4   Influence Based on CP-Nets

Most previous work either studied the influence among multiple agents while making decisions on a single issue (usually, in the context of social networks) [11, 14, 17–24, 37, 55, 58] or studied the dependency among multiple issues decided by a single agent (typically, using the model of CP-nets, i.e. conditional preference networks) [3, 4, 26, 56, 62, 63].

2.4.1   CP-Nets

CP-nets is a tool for representing and reasoning with conditional ceteris paribus preference statements [3], which is appropriate to describe multi-issue decision-making. In CP-nets, we should distinguish between (definitive) preference statements and conditional preference statements.

Example 2.2 (Western Restaurant Ordering Issue Represented by CP-nets) Image you are ordering dinner at a Western restaurant, which is a typical multi-issue decision-making [57], as you need to make decisions on a series of issues including main course, starter (appetizer), and drinks, etc. Assume there are three alternatives for the main course: {beef, fish, chicken}, there are two alternatives for the starter: {salad, ham}, and there are two alternatives for the drinks: {red wine, white wine}.

There are usually dependencies among multiple issues for decision-making, and we usually make decisions on primary issue(s) first and then on secondary issues, particularly considering that later decisions are usually affected by or subject to prior decisions [26] (it is better to let secondary decisions be subject to primary decisions than to let primary decisions be subject to secondary decisions). Although the starter will be delivered before the main course, the main course is undoubtedly more important than the starter for dinner.

---

[53] There are at least $C_m^2 = \frac{m(m-1)}{2}$ pairwise comparisons to be each assigned a value if the two values representing the preference of one alternative over the other and the preference of the other alternative over the first, respectively, are symmetric, e.g. sum to 1.

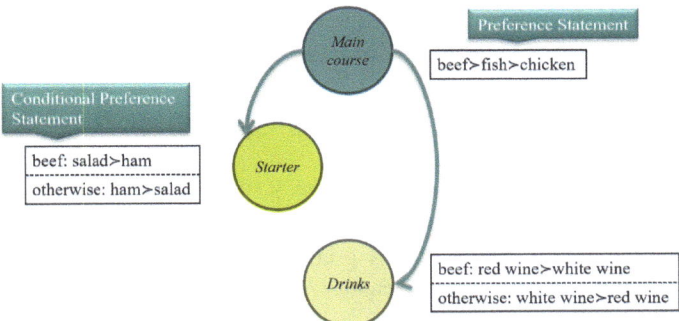

Fig. 2.5 A demonstration of CP-nets: Ordering in a Western restaurant. *Notes* The links represent the dependencies among issues

Assume your preference on the main course is: beef≻fish≻chicken, which is a sole and definitive preference and can be recorded by a preference statement. However, regarding your preferences on the starter and drinks, the situations are more complicated. It happens that your preference on a secondary issue depends on your choice on a primary issue. For instance, if you choose beefsteak as your main course, then you prefer salad to ham as your starter, because both beefsteak and ham are red meat and you would like to combine beef with some vegetables; while if you choose fish or chicken as your main course, then you prefer ham to salad as your starter, because you would like to eat some red meat besides while meat. Furthermore, if you ask a French: "Do you prefer red wine or white wine?" It may be hard for him or her to give a clear answer, because it depends. For instance, if you choose beefsteak as your main course, then you prefer red wine to white wine as your drinks; while if you choose fish or chicken as your main course, then you prefer white wine to red wine as drinks, because there is a saying that "red wine goes with red meat and white wine goes with white meat". Therefore, your preferences on the starter and drinks are not definitive but rather conditional, which depends on your choice on the main course, such preferences need to be recorded by conditional preference statements (Fig. 2.5).

2.4.2 Combining CP-Nets and Social Influence

A few studies (such as [48, 49]), through combining CP-nets and social influence, have put the influence among multiple agents and the dependency among multiple issues in the same model (influenced CP-nets), but they just discussed them separately,[54] and did not notice the influence across both multiple agents and multiple issues. If we think of multi-agent and multi-issue decision-making as a matrix, with agent as the horizontal axis and

---

[54] Influence among agents and dependency among issues are not "intertwined" in their model.

issue as the vertical axis,[55] this means that there are (horizontal) influences among multiple agents making decisions on a single issue and (vertical) dependencies among multiple issues (specifically, of later issues on former issues) decided by a single agent, but there is no (diagonal) influence across both multiple agents and multiple issues (as shown in Fig. 2.6) being discussed.

It should be noted that traditionally, dependency among issues is not deemed influence, but in fact, dependency among issues can be understood as a special kind of influence. Dependency among issues means that the preferences/choices of an agent on later issues will be affected by his or her own preferences/choices on former issues, in a sense, such "affected" just means "influenced".

Based on and improving upon the influenced CP-nets [48, 49], we provide a framework combining social networks and CP-nets[56] to model the influence across both multiple agents and multiple issues, where agents express their preferences regarding multiple issues as CP-nets, and influences (dependencies) among agents (issues) are represented by directed links in networks. Before formally presenting this framework, we introduce some previous work related to the study of influence in combinatorial and collective (multi-issue and multi-agent) decision-making with the help of Example 2.3, as follows:

Example 2.3 (United Nations Security Council Decision-making) This is a typical example of combinatorial and collective (multi-issue and multi-agent) decision-making with influences among agents (member states) and dependencies among issues (draft resolutions).

First, the UN Security Council decision-making is a typical multi-agent (collective) decision. There are 15 member states (including 5 permanent members and 10 nonpermanent members) collectively making decisions (voting) for each draft resolution. Each member state persuades (positively influences) its allies[57] and opposes (is negatively influenced by) its opponents,[58] in order to achieve desirable voting results on critical international issues and pursue its own state interests.[59] It is easy to see that the UN Security Council is full of various games, interactions and persuasions among member states, which means that there are influences (represented by the horizontal directed links shown in Fig. 2.6) among agents (member states). The models describing influence among agents are mainly social influence models [11, 14, 17–25, 34, 37, 55, 58].

Second, the UN Security Council decision-making is also a typical multi-issue (combinatorial) decision. The UN Security Council has made decisions on thousands of draft resolutions since the establishment of the United Nations. Moreover, there are many draft

---

[55] With entries each as the decision of an agent on an issue.

[56] Also involving social influence and social choice.

[57] To let them support what it supports, and oppose what it opposes.

[58] To support what they oppose, and oppose what they support.

[59] Or depress its opponents' interests.

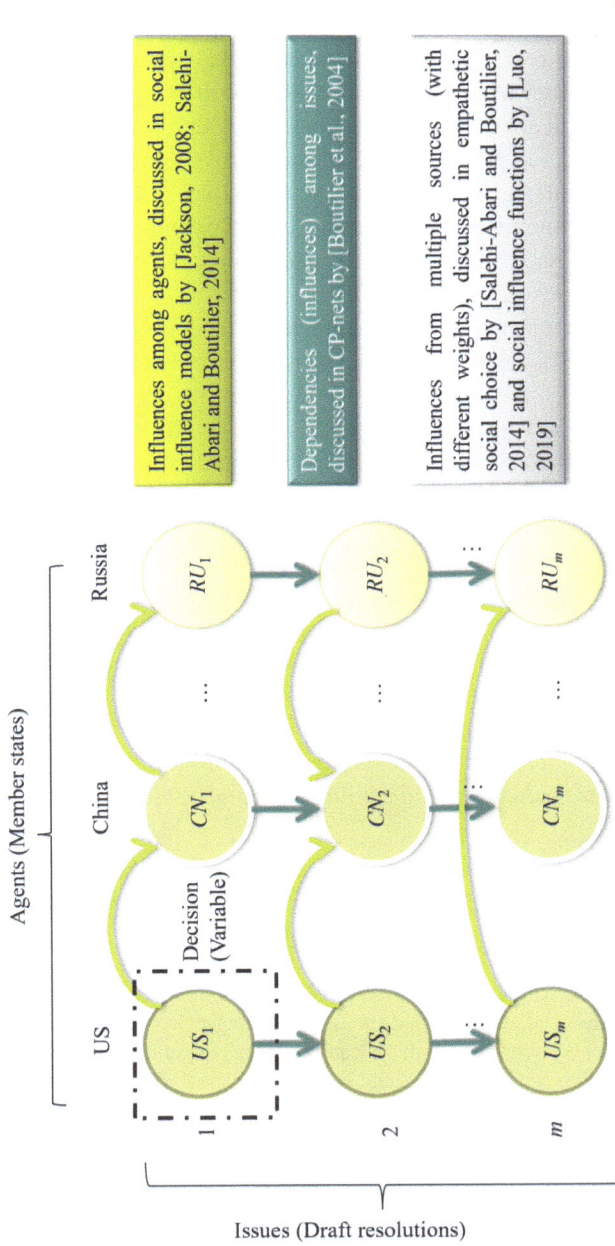

Fig. 2.6 The UN Security Council as a typical combinatorial and collective decision-making with influences among agents and dependencies among issues. *Notes* The yellow links represent the influences among multiple agents (member states) making decisions on a single issue (draft resolution), and the green links represent the dependencies (influences) among multiple issues (draft resolutions) decided by a single agent (member state)

resolutions frequently addressing the same subject (such as the Palestinian-Israeli issue[60] and the Syria issue). Usually, the votes of a given member state on later draft resolutions will be affected by or dependent on (refer to) its own votes on former draft resolutions with the same or similar subjects, which means that there are dependencies (represented by the vertical directed links shown in Fig. 2.6) among issues (draft resolutions). For instance, the United States has almost always used its veto power on draft resolutions against or unfavorable to Israel, not only for its own state interests, but also for its reputation in the international community, especially in the minds of its allies (a great power should be constant and trustworthy in its attitudes and behaviors on critical issues, providing stable expectations[61] for its allies and even for its opponents[62]). The typical model describing dependency among issues[63] is CP-nets [3, 4, 26, 56, 62, 63].

Third, it should be noted that the influences faced by a member state usually have multiple sources, i.e. a member state will be influenced by more than one member state at the same time, which complicates the process of influence and the determination of the result of influence.[64] For instance, a vote by China may be influenced by Russia, the United States and some other member states at the same time. As such, how do we address the multiple sources of influence and determine the resulting preference or choice, especially in light of opposite influencing directions (such as a positive influence from Russia and a negative influence from the United States) and diversified influencing strengths (such as a stronger influence from a great power and a weaker influence from a small country)? Models addressing multiple sources of influence include the empathetic social choice [58], *social influence functions* and *matrix influence function* discussed in this book, etc. The empathetic social choice [58] addressed multiple sources of influence in group decision-making in the context of a social network, setting a weight of influence of each influencing agent on each influenced agent, in which an agent's utility value for each alternative is collectively affected by both other agents' utilities and his or her own initial utility for the same alternative. As both the subject

---

[60] The resolutions on the Palestinian-Israeli issue, arms control and nuclear disarmament, the human rights issue occupy the majority of all resolutions adopted by the UN Security Council.

[61] Donald Trump may violate it.

[62] In order to manage potential conflicts.

[63] Or dependency among features of a multi-feature decision. For instance, when a family is choosing a car, they should make decisions on the budget, manufacturer (maker or brand), type, color, and other features. There are dependencies among these features. If the budget is set at 38,000 euros, then many manufacturers' cars will be out of question and should be eliminated from the manufacturer domain. If the chosen manufacturer is Ferrari®, then an SUV may not be an option and should be removed from the type domain.

[64] Influence is not as simple as we might expect, even for the influence just among multiple agents (while making decisions on a single issue), because the influence of reality faced by an agent usually comes not from a single agent at a time, but from more than one agent at the same time. For instance, when a family is choosing a car, if the preference of the husband is influenced by his wife, then a simple processing is to make the preference of the husband become identical with his wife's; however, if the husband is simultaneously influenced by his wife and children with different preferences, then how to determine the resulting preference is not that simple and straightforward.

of influence and the object of influence are utility values of agents, it is a typical cardinal (utility value-based) approach. Moreover, it assumes that all influence are positive, which may be an oversimplification compared to reality.[65]

However, previous work on influence models mainly studied how to address multiple sources of influence among agents' preferences or choices just on a single issue.[66] Overall, the influence from more than one agent making a decision on one issue (corresponding to the multiple influences in the horizontal dimension shown in Fig. 2.6) has been fully discussed (such as in the model of empathetic social choice [58]); besides, the dependency on (influence from) more than one issue for one agent's decision-making (corresponding to the multiple influences in the vertical dimension shown in Fig. 2.6) has been preliminarily described (such as in the model of CP-nets [26]); however, the influence across both more than one agent and more than one issue (which are multiple influences in the diagonal dimension) has been relatively ignored.

---

[65] In real-world settings, it is certain that there are positive influences from friends and family; however, it is impossible to avoid negative influences from enemies, opponents, or any person with a negative view of each other.

[66] Or on multiple issues independent of each other.

Part I

How to Address Multiple Sources of Influence in Group Decision-Making?

From social choice functions to *social influence functions* and from the KSB distance metric to a *matrix influence function*.

Social Influence Functions Based on Social Choice

We extend several classical social choice functions to signed and weighted *social influence functions (rules)* in the context of group decision-making with mutual influence, including *plurality influence rule, majority influence rule, Borda influence rule*, and *Condorcet influence rule*. Social choice functions take individual choices/preferences as input and give a collective choice/preference as output; likewise, *social influence functions* take influencing agents' choices/preferences as input and give the influenced agent's choice/preference as output. We provide an example of group decision-making with multiple sources of influence to apply and illustrate *social influence functions* we design.

Social influence functions can be understood as the mathematical expressions of *social influence rules*.

3.1   Social Influence Function: Nonordering Approach

In a nonranked choice context, every agent can choose only his or her first preferred alternative (candidate) but not present a full ordering among all alternatives (candidates) in his or her ballot. As agents can only observe other agents' choices about their most preferred alternatives, the influences among agents can only work in a nonordering way, flowing from the influencing agents' 1-of-m choices to the influenced agent's 1-of-m choice (assume there are m alternatives).[1] According to the example shown in Fig. 3.1, the influences flow from all influencing agents F_1, F_2, E, and M's 1-of-3 choices to agent M's 1-of-3 choice. As each agent's choice is made according to his or her first preferred alternative,[2] the influencing

---

[1] We cannot address the full preference orderings influencing and being influenced, because such information about the orderings among all alternatives is inaccessible (due to the ballot constraint and privacy).

[2] We do not consider strategic voting and manipulation here.

© The Author(s), under exclusive license to Springer Nature Switzerland AG 2026

H. Luo, *Influence Models in Group Decision-Making*, Synthesis Lectures on Computer Science, https://doi.org/10.1007/978-3-032-01352-1_3

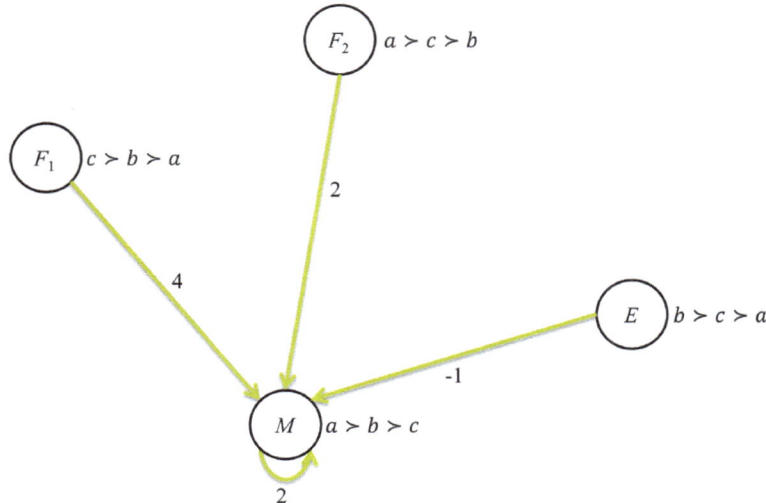

Fig. 3.1 An example of simultaneous influence of more than one agent

choice of agent F_1 will be c, the influencing choice of agent F_2 will be a, the influencing choice of agent E will be b, and the influencing choice of agent M will be a.

Example 3.1 (Multiple Sources of Influence) We use an example to display and compare how different approaches, including nonordering and ordering approaches, can be employed to address the simultaneous influence from more than one agent. As shown in Fig. 3.1, we assume there are four agents (named M, F_1, F_2, E) making choices among three alternatives (candidates) (a, b, c). We start from the perspective of agent M: agent M ("Me") is collectively influenced at the same time by agents F_1 (Friend 1), F_2 (Friend 2), E (Enemy), and M ("Myself") either, with respective preferences of $c \succ b \succ a, a \succ c \succ b, b \succ c \succ a$, and $a \succ b \succ c$ and with respective weights of influence of 4, 2, −1, and 2. Usually, a person is positively influenced by his or her friends (among friends, some may have closer relationships than others, differentiated by a larger or smaller value of positive weight[3]), negatively influenced by his or her enemies (positive influence means trying to be close to friends' preferences or choices and negative influence means trying to be far from enemies' preferences or choices), and positively influenced by (or referring to) his or her own former preference.

It is quite common that we will be influenced not only by others (friends, enemies) surrounding us but also by ourselves. My latter self will inevitably be influenced by my former self, and the weight of my own influence is usually positive. After all, there is a considerable degree of continuity and stability in human preference and behavior, especially

---

[3] Usually, the closer the friendship, the larger the weight of influence of a friend.

according to the principle of rationality. Only in extreme cases, such as when a person suffers serious setbacks and loses self-confidence, could his or her own influence be negative.[4] Such a setup of one's own influence can explain why some people are hard to be influenced by others while some other people are easy to be influenced by others, this is because the former individuals' own influences may have higher weights. For example, two persons share a common friend; the friend tries very hard to influence both of them, but only one of them is influenced and the other one insists because the weights of their own influences are different.

3.1.1   Plurality Influence Rule

Plurality rule (under which the alternative that obtains the highest number of votes wins) can be extended to a signed and weighted version referred to as a multi-influence aggregation method[5] that is used to measure the sum of the weighted influence scores[6] of each alternative on a particular (influenced) agent, obtained from all the influencing agents' choices.

Definition 3.1 (*Plurality Influence Score*) is the influence score of an alternative depending on how many times this alternative is first (most) preferred by all influencing agents and their respective weights of influence from the perspective of an influenced agent:

$$I_o^{\mathbf{P}}(i) = \sum_{O_1^j = o} w_{(j,i)} \quad i, j \in \mathbb{N}, o \in \mathbb{M}$$

where $I_o^{\mathbf{P}}(i)$ represents the plurality score of influence on agent i obtained by alternative o, O_1^j indicates the first preferred alternative of agent $j \in \mathbb{N}$, and $w_{(j,i)}$ is the weight of influence of agent j on agent i ($\mathbb{N}$ represents the set of agents and $\mathbb{M}$ represents the set of alternatives).

To compare the *plurality influence scores* among three alternatives a, b, c in the example shown in Fig. 3.1, the *plurality influence score* on agent M obtained by alternative a is $I_a^{\mathbf{P}}(M) = w_{(F_2,M)} + w_{(M,M)} = 2 + 2 = 4$ (2 from agent F_2 and 2 from agent M himself or

---

[4] Although in most cases a person's current preference will be positively influenced by (or dependent on) his or her own past preferences, while the negative influence is still reasonable in some cases: under a great blow, a person may feel very disappointed with his or her own past behaviors or unsatisfied with his or her current status, lose confidence in himself or herself, and be eager to make a difference, do something to improve the current situation, then one solution and reaction would be to refer more to other agents' preferences or choices, especially those of experts, authorities, winners, and less or even negatively to his or her own.

[5] Which can be used to address the simultaneous influence from more than one agent in group decision-making.

[6] In some sense, which can be understood as the weight of influence of each alternative (candidate) on the influenced agent.

herself), because both agents F_2 and M rank alternative a first and both of their weights of influence on agent M are 2; the *plurality influence score* on agent M obtained by alternative b is $I_b^{\mathbf{P}}(M) = w_{(E,M)} = -1$ (from agent E), because agent E ranks alternative b first and his or her weight of influence on agent M is -1; the *plurality influence score* on agent M obtained by alternative c is $I_c^{\mathbf{P}}(M) = w_{(F_1,M)} = 4$ (from agent F_1), because agent F_1 ranks alternative c first and his or her weight of influence on agent M is 4. Thus, both two alternatives a and c have the highest weighted score of influence (or, more briefly, the weight of influence) on agent M according to *plurality influence rule*; it is still uncertain[7] for the result of the influenced choice for agent M.

3.1.2 Majority Influence Rule

Majority rule (under which the alternative that obtains more than half of the votes wins) can also be extended to a signed and weighted version referred to as a multi-influence aggregation method that is used to measure the sum of the weighted influence scores[8] of each alternative on a particular (influenced) agent, obtained from all the influencing agents' choices.[9]

Definition 3.2 (*Majority Influence Score*) can be understood as the *plurality influence score* counted with one or more rounds, eliminating one or a portion of alternatives each round, until only one alternative obtains more than half of the total influence scores: $\frac{\sum_{j\in\mathbb{N}} w_{(j,i)}}{2}$. $I_o^{\mathbf{P}}(i)$ or $I_o^{\mathbf{M}}(i)$ can be used to represent the majority score of influence on agent i obtained by alternative o.

To compare the *majority influence scores* among three alternatives a, b, c in the example shown in Fig. 3.1, the *majority influence score* on agent M obtained by alternative a is $I_a^{\mathbf{P}}(M) = w_{(F_2,M)} + w_{(M,M)} = 2 + 2 = 4$, that obtained by alternative b is $I_b^{\mathbf{P}}(M) = w_{(E,M)} = -1$, and that obtained by alternative c is $I_c^{\mathbf{P}}(M) = w_{(F_1,M)} = 4$. Thus, both two alternatives a and c have the highest weighted score of influence (or, more briefly, the weight of influence) on agent M according to *majority influence rule*; it is still uncertain for the result of the influenced choice for agent M. Although the *majority influence score* on agent M obtained by alternative a and that obtained by alternative c are both 4 and both larger than half of the total influence scores: $\frac{4-1+4}{2} = \frac{7}{2}$, the influenced choice for agent M should be one alternative, which means we need to break the tie. Further, if we

---

[7] Both alternatives could be winners or we could use some other rules to determine one winner.

[8] In some sense, which can also be understood as the weight of influence of each alternative (candidate) on the influenced agent.

[9] Reference [24] also discussed a majority influence rule, but only in the context of binary (yes or no) decisions.

design a *two-round majority influence rule* (by extending the two-round system[10] to a multi-influence aggregation method), then only alternatives a and c (as the top 2) can be compared in the second round of influence competition (since alternative b as the bottom is eliminated in the first round): the *majority influence score* on agent M obtained by alternative a in the second round is $I_a^{\mathbf{P}}(M) = w_{(F_2,M)} + w_{(M,M)} = 2 + 2 = 4$ (2 from agent F_2 and 2 from agent M himself or herself), and that obtained by alternative c in the second round is $I_c^{\mathbf{P}}(M) = w_{(F_1,M)} + w_{(E,M)} = 4 - 1 = 3$ (4 from agent F_1 and -1 from agent E), because agent E's first preferred alternative b has been eliminated and agent E's second preferred alternative is c. Thus, alternative a has the highest weighted score of influence on agent M according to *two-round majority influence rule*, is the result of the influenced choice for agent M.

As we observe, the nonordering approach to address the simultaneous influence of more than one agent can be simple, while a critical drawback is: only the information about the most preferred alternatives is used, whereas all other preference information is ignored[11] (it is totally true under *plurality influence rule*; under *majority influence rule*, other preference information of a portion of agents can be used only after their first preferred alternatives have been eliminated[12]). However, all other preference information is also valuable; in particular, information about the least preferred alternatives is equally as valuable as that about the most preferred alternatives in some sense. In the process of multi-influence aggregation, if only the most preferred alternatives are collected but the least preferred alternatives are ignored,[13] the computed result of influence may be unreasonable.[14]

---

[10] If no alternative (candidate) can obtain more than half of the votes, then only the top 2 will advance to the second round. The two-round system is used in practice, such as in the French presidential election.

[11] For a group decision with n agents and m alternatives, an $n \times m$ (at least $n \times (m - 1)$) matrix is needed to display all preference information (in which the values in a given column indicate how an agent ranks all alternatives, and the values across each row indicate how all agents rank a particular alternative in a specific position in their preference orderings). However, in a nonordering approach, usually only a $1 \times n$ vector of preference information is used, and an $(m - 1) \times n$ matrix of preference information is ignored.

[12] Specifically, the information about the second preferred alternatives of a portion of agents will be used only after their first preferred alternatives have been eliminated, the information about the third preferred alternatives of a even smaller portion of agents will be used only after both their first and second preferred alternatives have been eliminated, and so forth.

[13] In some sense, the preference information that is closer to the top and that closer to the bottom are more important, both of which indicate stronger attitudes or feelings.

[14] Similar to the controversial winners produced by nonranked social choice methods in the examples [38] proposed by Charles Dodgson.

3.2   Social Influence Function: Ordering-Based Approach

In a ranked choice context, every agent can present a full ordering among all alternatives (candidates) in his or her ballot. As agents can observe other agents' full preference orderings over all the alternatives, the influences among agents can work in an ordering way, flowing from the influencing agents' preference orderings (there are $m! = m \times (m-1) \times \ldots \times 2 \times 1$ kinds[15] of possible orderings if there are m alternatives) to the influenced agent's preference ordering. As illustrated by the example shown in Fig. 3.1, the influences flow directly from all the influencing agents' preference orderings, including agent F_1's preference ordering $c \succ b \succ a$, agent F_2's preference ordering $a \succ c \succ b$, agent E's preference ordering $b \succ c \succ a$, and agent M's original preference ordering before the influence, $a \succ b \succ c$, to agent M's influenced preference ordering. To address the simultaneous influence of more than one agent in an ordering approach, new influence functions (rules) are needed.

3.2.1   Borda Influence Rule

Borda count (which usually assumes that a score of m is given to the first preferred alternative, a score of $m-1$ is given to the second preferred alternative, and so on until finally, a score of 1 is given to the least preferred alternative, i.e. allocating scores according to alternatives' ranks in preference orderings) can be extended to a signed and weighted version referred to as a multi-influence aggregation method that is used to measure the weighted sum of the influence scores of each alternative on a particular (influenced) agent, obtained from all the influencing agents' preference orderings.

Definition 3.3 (*Borda Influence Score*) is the influence score of an alternative depending on the ranks of this alternative in all influencing agents' preference orderings and their respective weights of influence from the perspective of an influenced agent:

$$I_o^{\mathbf{B}}(i) = \sum_{O_r^j = o} (m - r + 1) \times w_{(j,i)} \quad i, j \in \mathbb{N}, o \in \mathbb{M}$$

where $I_o^{\mathbf{B}}(i)$ represents the Borda score of influence on agent i obtained by alternative o, $r \in \{1, 2, \ldots, m\}$ represents the rank from 1st to mth, and O_r^j indicates the rth preferred (ranked) alternative by agent $j \in \mathbb{N}$.

To compare the *Borda influence scores* obtained by three alternatives a, b, c in the example shown in Fig. 3.1, considering there are three alternatives, it is assumed that a score

---

[15] For which alternative to rank first, there are m choices; for which alternative to rank second, there are $m-1$ choices left; and so on until finally, for which alternative to rank last, there is only 1 choice left.

of 3 is given to the first preferred alternative, a score of 2 is given to the second preferred alternative, and a score of 1 is given to the least preferred alternative:

(1) the Borda score of influence on agent M obtained by alternative a is $I_a^{\mathbf{B}}(M) = w_{(F_1,M)} \times 1 + w_{(F_2,M)} \times 3 + w_{(E,M)} \times 1 + w_{(M,M)} \times 3 = 4 \times 1 + 2 \times 3 + (-1) \times 1 + 2 \times 3 = 15$, in which 4×1 is the Borda score of influence on agent M of alternative a given by agent F_1, 4 is the weight of influence from agent F_1 to agent M, which is multiplied by 1 because a is least preferred by agent F_1; 2×3 is the Borda score of influence on agent M of alternative a given by agent F_2, 2 is the weight of influence from agent F_2 to agent M, which is multiplied by 3 because a is first preferred by agent F_2; -1×1 is the Borda score of influence on agent M of alternative a given by agent E, -1 is the weight of influence from agent E to agent M, which is multiplied by 1 because a is least preferred by agent E; and 2×3 is the Borda score of influence on agent M of alternative a given by agent M himself or herself, 2 is the weight of influence from agent M to agent M himself or herself, which is multiplied by 3 because a is first preferred by agent M himself or herself;

(2) the Borda score of influence on agent M obtained by alternative b is $I_b^{\mathbf{B}}(M) = w_{(F_1,M)} \times 2 + w_{(F_2,M)} \times 1 + w_{(E,M)} \times 3 + w_{(M,M)} \times 2 = 4 \times 2 + 2 \times 1 + (-1) \times 3 + 2 \times 2 = 11$, in which 4×2 is the Borda score of influence on agent M of alternative b given by agent F_1, 4 is the weight of influence from agent F_1 to agent M, which is multiplied by 2 because b is second preferred by agent F_1; 2×1 is the Borda score of influence on agent M of alternative b given by agent F_2, 2 is the weight of influence from agent F_2 to agent M, which is multiplied by 1 because b is least preferred by agent F_2; -1×3 is the Borda score of influence on agent M of alternative b given by agent E, -1 is the weight of influence from agent E to agent M, which is multiplied by 3 because b is first preferred by agent E; and 2×2 is the Borda score of influence on agent M of alternative b given by agent M himself or herself, 2 is the weight of influence from agent M to agent M himself or herself, which is multiplied by 2 because b is second preferred by agent M himself or herself;

(3) the Borda score of influence on agent M obtained by alternative c is $I_c^{\mathbf{B}}(M) = w_{(F_1,M)} \times 3 + w_{(F_2,M)} \times 2 + w_{(E,M)} \times 2 + w_{(M,M)} \times 1 = 4 \times 3 + 2 \times 2 + (-1) \times 2 + 2 \times 1 = 16$, in which 4×3 is the Borda score of influence on agent M of alternative c given by agent F_1, 4 is the weight of influence from agent F_1 to agent M, which is multiplied by 3 because c is first preferred by agent F_1; 2×2 is the Borda score of influence on agent M of alternative c given by agent F_2, 2 is the weight of influence from agent F_2 to agent M, which is multiplied by 2 because c is second preferred by agent F_2; -1×2 is the Borda score of influence on agent M of alternative c given by agent E, -1 is the weight of influence from agent E to agent M, which is multiplied by 2 because c is second preferred by agent E; and 2×1 is the Borda score of influence on agent M of alternative c given by agent M himself or herself, 2 is the weight of

influence from agent M to agent M himself or herself, which is multiplied by 1 because c is least preferred by agent M himself or herself.

In conclusion, c has the highest (weighted) Borda score of influence on agent M, a has the second highest score, and b has the lowest score; then, according to *Borda influence rule* we design (which works in a way similar to Borda count), the result of the influenced preference for agent M is $c \succ a \succ b$.

3.2.2   Condorcet Influence Rule

Condorcet method (which holds a pairwise comparison between each two alternatives and calls the alternative that can win over any other alternative in a pairwise comparison the Condorcet winner) can also be extended to a signed and weighted version referred to as a multi-influence aggregation method that is used to measure the weighted sum of the influence scores of each alternative on a particular (influenced) agent, obtained in every pairwise comparison with all other alternatives, based on all the influencing agents' preference orderings.

Definition 3.4 (*Condorcet Influence Score*) is the relative influence score of an alternative compared to another alternative depending on which alternative is preferred to the other in all influencing agents' preference orderings and their respective weights of influence from the perspective of an influenced agent:

$$I^{\mathbf{C}}_{o \succ o'}(i) = \sum_{o \succ_j o'} w_{(j,i)} \quad i, j \in \mathbb{N}, o, o' \in \mathbb{M}$$

where $I^{\mathbf{C}}_{o \succ o'}(i)$ represents the Condorcet score of influence on agent i obtained by alternative o compared with alternative o', and $o \succ_j o'$ indicates any agent ($j \in \mathbb{N}$) preferring alternative o to alternative o'.

To compare the pairwise *Condorcet influence scores* among three alternatives a, b, c in the example shown in Fig. 3.1, there will be $C_3^2 = \frac{3 \times 2}{2 \times 1} = 3$ pairs of comparison:

(1) a and b, the Condorcet score of influence on agent M obtained by alternative a compared with alternative b is $I^{\mathbf{C}}_{a \succ b}(M) = w_{(F_2,M)} + w_{(M,M)} = 2 + 2 = 4$ (in which the former 2 being here means that agent F_2 deems a better than b and the latter 2 being here means that agent M himself or herself also deems a better than b, and both their weights of influence on agent M himself or herself are 2), and the Condorcet score of influence on agent M obtained by alternative b compared with alternative a is $I^{\mathbf{C}}_{b \succ a}(M) = w_{(F_1,M)} + w_{(E,M)} = 4 - 1 = 3$ (in which the 4 being here means that agent F_1 deems b better

than a and the -1 being here means that agent E also deems b better than a, and their weights of influence on agent M are 4 and -1, respectively), as $I^{C}_{a \succ b}(M) > I^{C}_{b \succ a}(M)$, a has a stronger influence than b on agent M;

(2) a and c, the Condorcet score of influence on agent M obtained by alternative a compared with alternative c is $I^{C}_{a \succ c}(M) = w_{(F_2,M)} + w_{(M,M)} = 2 + 2 = 4$ (in which the former 2 being here means that agent F_2 deems a better than c and the latter 2 being here means that agent M himself or herself also deems a better than c, and both their weights of influence on agent M himself or herself are 2), and the Condorcet score of influence on agent M obtained by alternative c compared with alternative a is $I^{C}_{c \succ a}(M) = w_{(F_1,M)} + w_{(E,M)} = 4 - 1 = 3$ (in which the 4 being here means that agent F_1 deems c better than a and the -1 being here means that agent E also deems c better than a, and their weights of influence on agent M are 4 and -1, respectively), as $I^{C}_{a \succ c}(M) > I^{C}_{c \succ a}(M)$, a also has a stronger influence than c on agent M;

(3) b and c, the Condorcet score of influence on agent M obtained by alternative b compared with alternative c is $I^{C}_{b \succ c}(M) = w_{(E,M)} + w_{(M,M)} = -1 + 2 = 1$ (in which the -1 being here means that agent E deems b better than c and the 2 being here means that agent M himself or herself also deems b better than c, and their weights of influence on agent M himself or herself are -1 and 2, respectively), and the Condorcet score of influence on agent M obtained by alternative c compared with alternative b is $I^{C}_{c \succ b}(M) = w_{(F_1,M)} + w_{(F_2,M)} = 4 + 2 = 6$ (in which the 4 being here means that agent F_1 deems c better than b and the 2 being here means that agent F_2 also deems c better than b, and their weights of influence on agent M are 4 and 2, respectively), as $I^{C}_{b \succ c}(M) < I^{C}_{c \succ b}(M)$, c has a stronger influence than b on agent M.

In conclusion, according to *Condorcet influence rule* we design (which works in a way similar to Condorcet method), alternative a can be named the *Condorcet influence winner* for agent M, as it can beat any other alternative in a pairwise comparison of the (weighted) Condorcet scores of influence on agent M, and the preference ordering for agent M after being influenced is $a \succ c \succ b$ (alternative c loses to alternative a but wins over alternative b).[16]

---

[16] Fortunately, there is no *Condorcet influence paradox* here. This concept may have been first proposed in this book, which is similar to the Condorcet paradox. In such circumstance, we may need to extend Dodgson's method and Young's method, etc. into influence functions.

Matrix Influence Function Based on Ordering Matrix

4

We extend the KSB distance metric to a signed and weighted *matrix influence function* in the context of group decision-making with mutual influence. To achieve this, we define the rule of how to transform each preference ordering into an ordering matrix and set a distance metric to compute the distance between any two preference orderings (i.e. ordering matrices); then, the preference ordering that has the smallest weighted sum of distances from all influencing agents' preference orderings, among all the theoretically possible preference orderings (according to the set of alternatives), is the resulting preference of the influenced agent. This function takes influencing agents' preference orderings as input and give the influenced agent's preference ordering as output. We use the same example of group decision-making with multiple sources of influence in the previous chapter to apply and illustrate the *matrix influence function*.

4.1 Matrix Influence Function: Modeling

In real-world settings, it is quite common that an agent is simultaneously influenced by more than one agent in the process of group decision-making. Different influencing agents usually have different weights of influence, which can be stronger or weaker in strength, and positive or negative in polarity. To support the multi-influence aggregation in an ordering (ordinal preference) approach, we introduce a *matrix influence function*[1] that makes use of the classical KSB distance metric [2, 39, 62, 63] by extending it to a signed and weighed version. Bogart [2] generalizes the work by Kemeny and Snell [39] that obtains a distance

---

[1] Since this influence function requires preference orderings to be transformed into the corresponding matrices, it can be called the *matrix influence function*.

© The Author(s), under exclusive license to Springer Nature Switzerland AG 2026
H. Luo, *Influence Models in Group Decision-Making*, Synthesis Lectures on Computer Science, https://doi.org/10.1007/978-3-032-01352-1_4

measure on strict[2] partial orderings as the unique metric satisfying several natural axioms
[62]. This metric, called the KSB metric (named based on the initials of the three contributors
[2, 39]), is defined in terms of a matrix representation of preference orderings [62].

Definition 4.1 (*Group Decision-making Society with Mutual-Influence (1)*) Assume a soci-
ety $\mathbb{S} = \{\mathbb{N}, \mathbb{M}, \mathbb{P}, \mathbb{W}\}$, where $\mathbb{N} = \{1, 2, ..., n\}$ is the set of all agents (a general term that
can represent a person or an artificial intelligence in nature and that can represent a decision-
maker, a voter, a game player, etc. in function); $\mathbb{M} = \{o_1, o_2, ..., o_m\}$ is the set of all alternatives (candidates); $\mathbb{P} = \{P_{(1)}, P_{(2)}, ..., P_{(n)}\}$ is the set of all agents' preferences; all possible
preference orderings according to the set of alternatives $\mathbb{M}$ include $m!$ kinds (as the number of alternatives is m), the set of all possible preference orderings can thus be denoted
as $\mathbb{P}[\mathbb{M}] = \{p_1, p_2, ..., p_{m!}\}$; $\mathbb{W}$ is the matrix whose entries are the weights of influence
between each two agents, $\mathbb{W} = [w_{(i,j)}]$ $(i, j \in \mathbb{N})$, in which $w_{(i,j)}$ means the weight of
influence from agent i to agent j, the weight value indicates both the strength and polarity
of the influence, $w_{(i,j)} > 0$ means a positive influence, $w_{(i,j)} < 0$ means a negative influ-
ence, $w_{(i,j)} = 0$ means there is no influence from agent i to agent j, and the higher $|w_{(i,j)}|$,
the stronger the influence from agent i to agent j.

It should be noted that the weights of influence between two agents may not be symmetric,
i.e. $w_{(i,j)} \neq w_{(j,i)}$. For instance, a common real-world scenario is that you think someone
is your intimate friend (corresponding to a stronger influence), but he or she thinks of you
as just an ordinary friend (corresponding to a weaker influence); or even worse, you think
someone is your friend (corresponding to a positive influence), but he or she does not think
similarly and thinks of you as a bother (corresponding to a negative influence).

Definition 4.2 (*Ordering Matrix*) is a matrix transformed from a preference ordering, which
can be written as $OM = [om_{o,o'}]$, in which $om_{o,o'}$ is the comparison between two alternatives o, o' ($o, o' \in \mathbb{M} = \{o_1, o_2, ..., o_m\}$). An ordering matrix is in canonical form if the
column and row are ordered lexicographically with the set of alternatives:

$$
OM = \begin{matrix} o_1 \\ o_2 \\ \cdots \\ o_m \end{matrix}
\begin{pmatrix}
om_{o_1,o_1} & om_{o_2,o_1} & \cdots & om_{o_m,o_1} \\
om_{o_1,o_2} & om_{o_2,o_2} & \cdots & om_{o_m,o_2} \\
\cdots & \cdots & \cdots & \cdots \\
om_{o_1,o_m} & om_{o_2,o_m} & \cdots & om_{o_m,o_m}
\end{pmatrix}
$$

---

[2] "Strict" means that for any two alternatives (candidates), we can distinguish which one is better.

For the preference ordering of agent i, $P_{(i)}$, its corresponding ordering matrix is $OMP_{(i)} = [om_{o,o'}^{P_{(i)}}]$:

$$om_{o,o'}^{P_{(i)}} = \begin{cases} 1 & \text{if } o \text{ is strictly preferred to } o' \text{ by } P_{(i)} \\ -1 & \text{if } o' \text{ is strictly preferred to } o \text{ by } P_{(i)} \quad i \in \mathbb{N} \\ 0 & \text{if } o \text{ is just } o' \text{ itself} \end{cases}$$

In the above, we assume that there is a strict preference[3] [32] ordering among all alternatives for every agent.

Definition 4.3 (*Distance between Preference Orderings*) is a distance in the sense of the minimum number of swapping adjacent alternatives to make two preference orderings identical or the number of pairwise disagreements between two preference orderings. The larger the number of pairwise disagreements between two preference orderings, the longer the distance between them. Assume there are two agents i and j with respective preference orderings $P_{(i)}$ and $P_{(j)}$; let $om_{o,o'}^{P_{(i)}}$ and $om_{o,o'}^{P_{(j)}}$ be corresponding entries from respective ordering matrices $OMP_{(i)}$ and $OMP_{(j)}$; then, the distance between preference orderings $P_{(i)}$ and $P_{(j)}$ is (define Dis as the distance function):

$$Dis(P_{(i)}, P_{(j)}) = \sum_{o \in \mathbb{M}} \sum_{o' \in \mathbb{M}} |om_{o,o'}^{P_{(i)}} - om_{o,o'}^{P_{(j)}}| \quad i, j \in \mathbb{N}$$

Definition 4.4 (*Matrix Influence Score*) is the influence score of a preference ordering depending on the weighted sum of its distances from all influencing agents' preference orderings from the perspective of an influenced agent, which is computed by firstly transforming all influencing preference orderings into corresponding ordering matrices, and then calculating the distance:

$$I_p^{\mathbf{M}}(i) = \sum_{j \in \mathbb{N}} w_{(j,i)} Dis(P_{(j)}, p) \quad i \in \mathbb{N}, p \in \mathbb{P}[\mathbb{M}]$$

where $I_p^{\mathbf{M}}(i)$ represents the matrix score of influence on agent i obtained by preference ordering p (which can be any possible preference ordering).

---

[3] In fact, in a more complicated situation, there may be indifference [32] or incomparability between two alternatives (indifference and incomparability should be distinguished, as two alternatives that are indifferent to each other are still comparable), or even lack of (i.e. missing) relevant information about the preference between two alternatives, all of which make a strict order infeasible, constituting "otherwise" except strict preference.

Here, the score of influence on a particular (influenced) agent obtained by a preference ordering is the weighted sum of its distance from all the influencing agents' preference orderings, thus the lower the score[4] (i.e. the shorter the distance), the more likely it is to be the resulting preference for the influenced agent.

Definition 4.5 (*Matrix Influence Rule*) Let $P'_{(i)}$ be the preference of agent i after being influenced; the result of the influenced preference is one possible preference ordering ($p \in \mathbb{P}[\mathbb{M}]$) which has the minimum weighted sum of distances from all influencing preferences, compared with all other possible preference orderings:

$$P'_{(i)} = \arg \min_{p \in \mathbb{P}[\mathbb{M}]} [\sum_{j \in \mathbb{N}} w_{(j,i)} Dis(P_{(j)}, p)] \quad i \in \mathbb{N}$$

If this influence function considers the iteration of influence and multiperiod interactions (agents influence mutually at each interaction), let $P_{(i)}(t + 1)$ be the preference of agent i after tth mutual influence, and let $P_{(j)}(t)$ be the preference of agent j at tth mutual influence, then:

$$P_{(i)}(t + 1) = \arg \min_{p \in \mathbb{P}[\mathbb{M}]} [\sum_{j \in \mathbb{N}} w_{(j,i)} Dis(P_{(j)}(t), p)] \quad i \in \mathbb{N}$$

Since the weight of influence can be positive or negative (such as a positive influence from a friend or a negative influence from an enemy), it will partly play a role in finding the "closest" possible preference ordering in distance from the positively influencing agents' preferences and partly play a role in finding the "farthest" possible preference ordering in distance from the negatively influencing agents' preferences. This process will be demonstrated explicitly by finding the result of the influenced preference according to the example shown in Fig. 3.1.

4.2   Matrix Influence Function: Application

The *matrix influence function* is a signed and weighted version of the KSB metric [2, 39, 62, 63] in the context of group decision-making with multiple influences; using this function, the feasible (possible) preference ordering that has the minimum weighted sum of distances from all the influencing agents' preference orderings compared with all the other feasible preference orderings should be chosen as the resulting preference for an influenced agent. Different from *social influence functions* we built (such as *Borda influence function* and *Condorcet influence function*) that focus on comparing individual alternatives (such as finding which alternative has the highest score of influence, or which alternative can beat any other alternative in a pairwise comparison of the influence scores), the *matrix influence function* directly handles full preference orderings among all alternatives.

---

[4] Which is opposite to *social influence functions*, where the higher the score, the better.

Table 4.1 The ordering matrices of all feasible preference orderings over three alternatives a, b, c

$$OM_{a \succ b \succ c} = OM_{P_{(M)}} = \begin{array}{c} \\ a \\ b \\ c \end{array} \begin{pmatrix} a & b & c \\ 0 & -1 & -1 \\ 1 & 0 & -1 \\ 1 & 1 & 0 \end{pmatrix}$$

$$OM_{a \succ c \succ b} = OM_{P_{(F_2)}} = \begin{array}{c} \\ a \\ b \\ c \end{array} \begin{pmatrix} a & b & c \\ 0 & -1 & -1 \\ 1 & 0 & 1 \\ 1 & -1 & 0 \end{pmatrix}$$

$$OM_{b \succ a \succ c} = \begin{array}{c} \\ a \\ b \\ c \end{array} \begin{pmatrix} a & b & c \\ 0 & 1 & -1 \\ -1 & 0 & -1 \\ 1 & 1 & 0 \end{pmatrix}$$

$$OM_{b \succ c \succ a} = OM_{P_{(E)}} = \begin{array}{c} \\ a \\ b \\ c \end{array} \begin{pmatrix} a & b & c \\ 0 & 1 & 1 \\ -1 & 0 & -1 \\ -1 & 1 & 0 \end{pmatrix}$$

$$OM_{c \succ a \succ b} = \begin{array}{c} \\ a \\ b \\ c \end{array} \begin{pmatrix} a & b & c \\ 0 & -1 & 1 \\ 1 & 0 & 1 \\ -1 & -1 & 0 \end{pmatrix}$$

$$OM_{c \succ b \succ a} = OM_{P_{(F_1)}} = \begin{array}{c} \\ a \\ b \\ c \end{array} \begin{pmatrix} a & b & c \\ 0 & 1 & 1 \\ -1 & 0 & 1 \\ -1 & -1 & 0 \end{pmatrix}$$

Notes In which $a \succ b \succ c, a \succ c \succ b, b \succ c \succ a$, and $c \succ b \succ a$ are preference orderings that are held by the four agents M, F_2, E, and F_1, respectively, in the example shown in Fig. 3.1, and $b \succ a \succ c$ and $c \succ a \succ b$ are preference orderings that are not held by anyone but are feasible theoretically

To compute the distance between any two preference orderings, first, the ordering matrix OM for each feasible preference ordering (regardless of whether it is possessed by an agent in the group already or just exists theoretically) should be given, as shown in Table 4.1 according to the example shown in Fig. 3.1.

Further, based on the *matrix influence rule*, it finds one possible preference ordering p ($p \in \{a \succ b \succ c, a \succ c \succ b, b \succ a \succ c, b \succ c \succ a, c \succ a \succ b, c \succ b \succ a\}$) that has the minimum weighted sum of distances from all the influencing agents' preference orderings, including $P_{(F_1)}, P_{(F_2)}, P_{(E)}$, and $P_{(M)}$ (the weights of influence from agents F_1, F_2, E, and M to agent M himself or herself are 4, 2, −1, and 2, respectively):

Table 4.2 Finding the result of the influenced preference by the *matrix influence rule*

	$c \succ_{F_1} b \succ_{F_1} a$ (4)	$a \succ_{F_2} c \succ_{F_2} b$ (2)	$b \succ_E c \succ_E a$ (−1)	$a \succ_M b \succ_M c$ (2)	Dis
$a \succ b \succ c$	12×4	4×2	8×-1	0×2	48
$a \succ c \succ b$	8×4	0×2	12×-1	4×2	28
$b \succ a \succ c$	8×4	8×2	4×-1	4×2	52
$b \succ c \succ a$	4×4	12×2	0×-1	8×2	56
$c \succ a \succ b$	4×4	4×2	8×-1	8×2	32
$c \succ b \succ a$	0×4	8×2	4×-1	12×2	36

Notes The 4 preference orderings in the horizontal axis are the four influencing agent's preference orderings, $P_{(F_1)}$, $P_{(F_2)}$, $P_{(E)}$, and $P_{(M)}$, respectively (the numbers in parentheses after the preference orderings are their respective weights of influence); the 6 preference orderings in the vertical axis are all theoretically feasible preference orderings, all of which can be the result of the influenced preference for agent M. For every entry in each column, the first number is the distance between two preference orderings with one in the horizontal axis and the other in the vertical axis, the second number is the weight of influence of the influencing (agent's) preference, the two numbers multiplied together give us a weighted distance. We sum the weighted distances in each row to obtain the weighted sum of distances of each feasible preference ordering from all the influencing agents' preference orderings, and find which one is the minimum

$$P'_{(M)} = \arg \min_{p}[4Dis(P_{(F_1)}, p) + 2Dis(P_{(F_2)}, p) - Dis(P_{(E)}, p) + 2Dis(P_{(M)}, p)]$$

If we consider the iteration of influence and multiperiod interactions (among agents), let $P_{(M)}(t + 1)$ be the preference of agent M after tth mutual influence, and let $P_{(M)}(t)$ be the preference of agent M at tth mutual influence, then:

$$P_{(M)}(t + 1) = \arg \min_{p}[4Dis(P_{(F_1)}(t), p) + 2Dis(P_{(F_2)}(t), p) - Dis(P_{(E)}(t), p) + 2Dis(P_{(M)}(t), p)]$$

The influences are diversified in strength (stronger or weaker) and opposite in polarity (positive or negative).[5] There are both positive influences from friends and negative influences from enemies. Thus, the distance we compute is not only weighted but also signed. It will play a role in finding the "closest" preference ordering that is possible while computing the distance from positively influencing preferences and finding the "farthest" preference ordering that is possible while computing the distance from negatively influencing preferences.

According to the example shown in Fig. 3.1 and from the perspective of agent M being influenced, the computation outcome of the weighted sums of distances of all 6 possible preference orderings from all the influencing agents' preference orderings is shown in Table 4.2, and the result of the influenced preference for agent M is $a \succ c \succ b$, which possesses the minimum weighted sum of distances: $8 \times 4 + 0 \times 2 + 12 \times -1 + 4 \times 2 = 28$.

---

[5] i.e. different influencing agents have various strengths of influence and opposite polarities of influence.

In conclusion, how to address the simultaneous influence of more than one agent on another agent in group decision-making by both nonordering and ordering approaches have been demonstrated. We extend classic social choice functions, such as plurality rule, majority rule, Borda count and Condorcet method, to signed and weighted *social influence functions* in the context of a social network, where the influence represented by directed links among agents' choices or preferences can vary both in strength: stronger or weaker, and in polarity: positive or negative. Moreover, we extend the KSB distance metric to a signed and weighted *matrix influence function*: we define the rule of how to transform each preference ordering into an ordering matrix (using a matrix to represent each preference ordering) and set a distance metric to compute the distance between any two preference orderings (i.e. ordering matrices); then, the theoretically feasible preference ordering that has the minimum weighted sum of distances from all influencing agents' preferences is the resulting preference of the influenced agent.

Part II
Individual, Coalitional and Structural Influence in Group Decision-Making

Influences from independent agents, from coalitional agents, and from structured agents.

Graphical and Mathematical Expressions of the Three Levels of Influence

We formally present the *three levels of influence (individual, coalitional and structural influence)* in group decision-making using both graphs and functions.

Most previous work discussed the influence from multiple agents but mainly in an individual (independent) way, which assumed that all influencing agents exert their own influences independently from each other. Thus, the resulting preference[1] or choice for an influenced agent could be a simple linear weighted aggregation of the preferences or choices of all the influencing agents. However, these previous studies have less discussed coalitions[2] of influencing agents (possessing the same or similar beliefs, opinions, or making the same choices) and, in particular, have largely ignored that structures (i.e. influencing relationships) within influencing agents can also produce additional and extraordinary influencing effects.

If we look back at the previous work on influence models in group decision-making [11, 13, 14, 17–25, 34, 37, 42, 43, 45, 48, 49, 55, 58], we could conclude that the discussions about the influence from coalitions (of agents) and particularly the influence from structures (among agents) actually advance and complete the theoretical system of influence study in group decision-making.[3] At this stage, we could present a new analytical framework of influence called the *three levels of influence*, in which the influences discussed in the previous work can be mainly classified as the *level I influence from independent agents* (called *individual influence*), while the *level II influence from coalitional agents* (called *coalitional influence*) has been less discussed and the *level III influence from structured agents* (called *structural influence*) has been nearly ignored. Only in the third level, the

---

[1] Including utility, belief, and opinion, etc.

[2] References [20–24] discussed the influence of a coalition.

[3] Which also indicates the place of this work in the literature.

© The Author(s), under exclusive license to Springer Nature Switzerland AG 2026
H. Luo, *Influence Models in Group Decision-Making*, Synthesis Lectures on Computer Science, https://doi.org/10.1007/978-3-032-01352-1_5

influence from structures (i.e. influencing relationships[4]) themselves are considered in the process and result computation of multiple influences, and structures are regarded as origins of influence, not just as paths or channels of influence anymore.[5]

Definition 5.1 (*Group Decision-making Society with Mutual-Influence (2)*) Assume a society $\mathbb{S} = \{\mathbb{N}, \mathbb{M}, \mathbb{P}, \mathbb{C}, \mathbb{W}\}$, where $\mathbb{N} = \{1, 2, .., n\}$ is the set of all agents (a general term that can represent a person or an artificial intelligence in nature and that can represent a decision-maker, a voter, a game player, etc. in function); $\mathbb{M} = \{o_1, o_2, ..., o_m\}$ is the set of all alternatives (candidates); $\mathbb{P} = \{P_{(1)}, P_{(2)}, ..., P_{(n)}\}$ is the set of all agents' preferences (such as utilities, beliefs, decision-making probabilities); $\mathbb{C} = \{C_{(1)}, C_{(2)}, ..., C_{(n)}\}$ is the set of all agents' choices out of the set of alternatives; $\mathbb{W}$ is the matrix whose entries are the weights of influence between each two agents, $\mathbb{W} = [w_{(i,j)}]$ ($i, j \in \mathbb{N}$), in which $w_{(i,j)}$ means the weight of influence from agent i to agent j.

5.1   Level I Influence from Independent Agents

Definition 5.2 (*Individual Influence*) is the influence from agents as independent (separate) individuals. When an agent is simultaneously influenced by more than one agent but these influencing agents are independent from each other, separately exerting their own influences, the influences from different agents can be simply linearly summed by their respective weights (of influence). This kind of influence from independent agents can be expressed by a directed link marked as $\rightarrow_{x,y}$, in which x indicates the influencing subject and y indicates the influenced object. The weight of this kind of influence can be defined as $w_{x,y}$, which indicates the weight of influence from agent x to agent y.

Example 5.1 (A Graphical Expression of Individual Influence) As shown in Fig. 5.1, we assume that there are five agents 1–5, where agent 5 is simultaneously influenced by agents 1, 2, 3, 4, and 5 (himself or herself[6]), who all exert their own influences separately on agent 5. This is the basic assumption of most previous work assumed so far. Thus, there are five *individual (independent) influences*, which are from agent 1 to agent 5 (marked as $\rightarrow_{1,5}$), from agent 2 to agent 5 (marked as $\rightarrow_{2,5}$), from agent 3 to agent 5 (marked as $\rightarrow_{3,5}$), from agent 4 to agent 5 (marked as $\rightarrow_{4,5}$), and from agent 5 to agent 5 himself or herself (marked as $\rightarrow_{5,5}$).

---

[4] Among influencing agents.

[5] In another word, *structural influence* is the influence (originating) from structures rather than (passing) through structures.

[6] As mentioned earlier, the influence from oneself (self influence) cannot be ignored. In reality, your current preference or choice on an issue is remarkably affected by your previous preferences or choices on the same or similar issues, which explains well why, under identical influences from other people, some people can insist on their own preferences or choices while others change.

Fig. 5.1 Level I-influence from individual (independent) agents

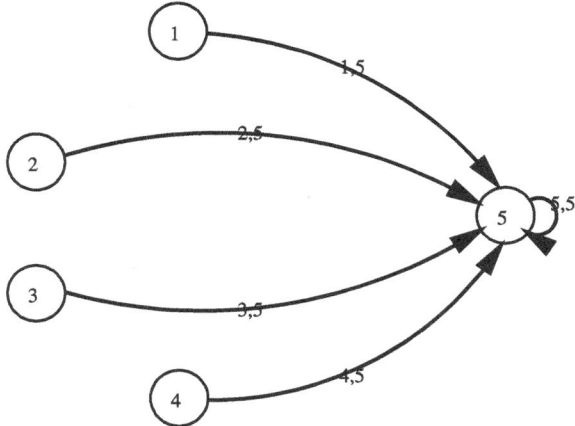

If the agents' preferences (such as utilities, beliefs) about an issue are expressed in terms of a unidimensional value, then the resulting preference of the influenced agent is a simple weighted average of all the influencing agents' preferences:

$$P_5' = \frac{w_{1,5}P_1 + w_{2,5}P_2 + w_{3,5}P_3 + w_{4,5}P_4 + w_{5,5}P_5}{|w_{1,5}| + |w_{2,5}| + |w_{3,5}| + |w_{4,5}| + |w_{5,5}|}$$

Definition 5.3 (*Individual Influence Function*) transforms multiple independent influencing agents' preferences or choices to the influenced agent's preference or choice. If this influence function takes into account the multiperiod interactions among agents and the iteration of mutual influence, then:

$$P_{(i)}(t+1) = \frac{\sum_{j\in\mathbb{N}} w_{(j,i)} P_{(j)}(t)}{\sum_{j\in\mathbb{N}} |w_{(j,i)}|} \quad i \in \mathbb{N}$$

In which $P_{(i)}(t+1)$ represents the preference of agent i after tth mutual influence, $P_{(j)}(t)$ represents the preference of agent j at tth mutual influence, and $w_{(j,i)}$ represents the weight of influence from agent j to agent i.

A more complex setting is that the weight of influence varies with time (becoming stronger or weaker in strength, and changing from positive to negative or vice versa in polarity),[7] then it should be expressed as $w_{(j,i)}(t)$, representing the weight of influence from agent j to agent i at tth mutual influence. For example, as time goes on, you may trust your friends

---

[7] Reference [42] designed a model combining agent-based modeling and simulation, decision theory, social network and system dynamics, where the (voting) preference of each agent will be influenced by the preferences of all other agents with the weight of influence $\neq 0$, and the weights of influence of different influencing agents from the perspective of the influenced agent will be affected by the voting results and his or her expectation psychology and comparison psychology.

more or less, hate your enemies more or less, make friends with your enemies or become enemies with your friends, for various reasons.

5.2   Level II Influence from Coalitional Agents

Definition 5.4 (*Coalitional Influence*) considers (a portion of) influencing agents as a coalition (united group), especially those agents possessing the same or similar beliefs, opinions (like an opinion alliance), or making the same choice (like a voting coalition), etc., which, under certain circumstances, will create an extra influencing effect (called *coalitional influence*) on the influenced agent. Thus, the multiple influences felt by the influenced agent may not merely be a simple weighted sum of all influences from individuals. This *coalitional influence* can be expressed by a dashed directed link marked as $--\!\!\!\rightarrow_{\{x_1, x_2, ..., x_c\}, y}$, where $\{x_1, x_2, ..., x_c\}$ is a coalition of influencing agents holding the same or similar preferences or making the same choice and y is the influenced agent. The weight of this kind of influence can be defined as $w_{\{x_1, x_2, ..., x_c\}, y}$, which indicates the weight of influence from the coalition of agents $\{x_1, x_2, ..., x_c\}$ to agent y.

A daily example is provided to illustrate why we should consider *coalitional influence*.[8]

Example 5.2 (Paper Submission and Review) When you submit a paper to a computer science conference and receive three reviews, if only one or two reviewers give negative evaluations, this would not seriously damage your confidence and feelings, and you might think that they do not really understand your paper, but if three independent reviewers all rate your paper as "rubbish", then you would probably feel despairing (the hard feeling would be much more than three times as strong as a single negative evaluation), which can be understood as a kind of coalitional influencing effect produced by multiple influencing agents having the same or similar views.

Example 5.3 (A Graphical Expression of Coalitional Influence) As shown in Fig. 5.2, again we assume that there are five agents 1–5, where agent 5 is simultaneously influenced by agents 1, 2, 3, 4, and 5 (himself or herself), but agent 5 further observes or believes that agents 1, 2, and 3 have the same or similar preferences (beliefs, opinions, etc.). Then, not only do the five agents 1, 2, 3, 4, and 5 all separately exert their own influences on agent 5, but also the coalition formed by agents 1, 2, and 3 (as an opinion alliance) would exert an extra *coalitional influence* on agent 5, which is marked as $--\!\!\!\rightarrow_{\{1,2,3\},5}$.

If the agents' preferences (such as utilities, beliefs) about an issue are expressed in terms of a unidimensional value, let $w_{\{1,2,3\},5}$ represents the weight of influence from the set of agents

---

[8] A similar concept is peer pressure. Peer pressure is the influence exerted by a peer group or individuals that encourage others to change their attitudes, values or behaviors to conform to group norms [50].

Fig. 5.2 Level II-influence
from coalitional agents

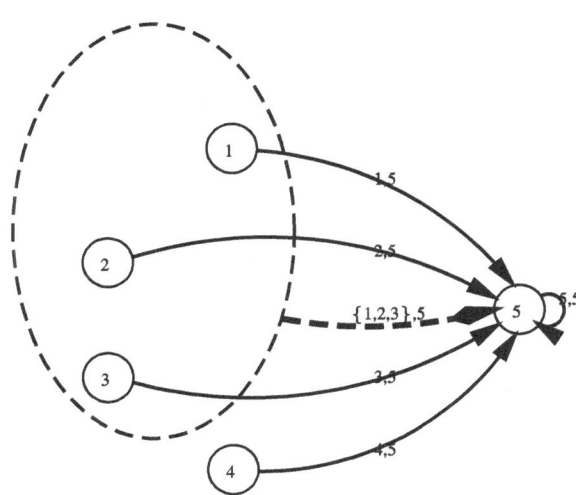

{1, 2, 3} as a coalition (ensemble) to agent 5, and define χ as the function for *coalitional influence*, which uses the preferences of the agents forming a coalition as input, then:

$$P_5' = \frac{w_{1,5}P_1 + w_{2,5}P_2 + w_{3,5}P_3 + w_{4,5}P_4 + w_{5,5}P_5 + w_{\{1,2,3\},5}\chi\{P_1, P_2, P_3\}}{|w_{1,5}| + |w_{2,5}| + |w_{3,5}| + |w_{4,5}| + |w_{5,5}| + |w_{\{1,2,3\},5}|}$$

Definition 5.5 (*Coalitional Influence Function*) transforms multiple coalitional influencing agents' preferences or choices to the influenced agent's preference or choice. If this influence function is expressed in a general multiperiod form (taking into account the iteration of mutual influence), then:

$$P_{(i)}(t+1) = \frac{\sum_{j\in\mathbb{N}} w_{(j,i)}P_{(j)}(t) + \sum_{c\in\mathbb{C}_{[\mathbb{N}](i)}(t)} w_{(c,i)}\chi[c]}{\sum_{j\in\mathbb{N}} |w_{(j,i)}| + \sum_{c\in\mathbb{C}_{[\mathbb{N}](i)}(t)} |w_{(c,i)}|} \quad i \in \mathbb{N}$$

In which $\mathbb{C}_{[\mathbb{N}](i)}(t)$ represents the set of all coalitions of influencing agents with the same or similar preferences within the set of all agents $\mathbb{N}$ from the perspective of agent i at tth mutual influence, $\chi[c]$ ($c \in \mathbb{C}_{[\mathbb{N}](i)}(t)$) represents the *coalitional influence* produced by coalition c, and $w_{(c,i)}$ represents the weight of influence from coalition c to agent i.

The above is just a general mathematical expression of how to address *coalitional influence*, a specific *coalitional influence function* can be found in the next chapter.

5.3   Level III Influence from Structured Agents

Definition 5.6 (*Structural Influence*) comes from the influencing relationships among influencing agents. The structures constituted by influencing relationships, which are represented by links in a social network, can also be perceived as origins of influence but not just paths of influence, and will produce an extra influencing effect (called *structural influence*) on the influenced agent. This *structural influence* can be expressed by a dashed directed link marked as $--\!\!\rightarrow_{x_1 \rightarrow x_2, y}$. The influencing subject discussed here is not an individual x_1 or x_2 or a coalition of x_1 and x_2 but rather an influencing relationship between x_1 and x_2 (specifically, from x_1 to x_2), and the influenced object is individual y. The weight of this kind of influence can be defined as $w_{x_1 x_2, y}$, which indicates the weight of influence from the influencing relationship (x_1 influenced x_2) to agent y.

The specific reasons why there should be *structural influence* considered in the process of multiple influences will be discussed in details in the next chapter. In one word, people may look down upon "followers"[9] and think highly of "leaders" (connected by influencing relationships).

Example 5.4 (A Graphical Expression of Structural Influence) As shown in Fig. 5.3, again we assume that there are five agents 1–5, where agent 5 is simultaneously influenced by agents 1, 2, 3, 4, and 5 (himself or herself), but agent 5 further finds or believes that among the three influencing agents 1, 2, and 3, there are two influencing relationships among them: one is from agent 1 to agent 2 and the other is from agent 1 to agent 3 (e.g. both agent 2 and agent 3 follow[10] agent 1). Thus, not only do the five agents 1, 2, 3, 4, and 5 all separately exert their own influences on agent 5, but also the two influencing relationships among them

Fig. 5.3 Level III-influence from structured agents

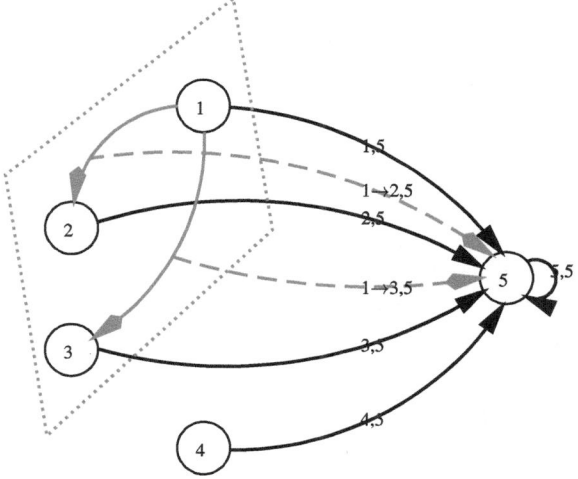

---

[9] "Following" is a specific form of influencing relationships.
[10] "Following" is a specific form of being influenced.

would produce extra *structural influences* on agent 5, which are marked as $--\!\!\rightarrow_{1\rightarrow2,5}$ and $--\!\!\rightarrow_{1\rightarrow3,5}$, respectively.

If the agents' preferences (such as utilities, beliefs) about an issue are expressed in terms of a unidimensional value, let $w_{12,5}$ and $w_{13,5}$ respectively represent the weights of influence from the influencing relationship $1 \rightarrow 2$ (agent 1 influenced agent 2) to agent 5 and from the influencing relationship $1 \rightarrow 3$ (agent 1 influenced agent 3) to agent 5, and define φ as the function for *structural influence*, which uses the preferences of the influencing agent and the influenced agent[11] connected by influencing relationships as input, then:

$$P_5' = \frac{w_{1,5}P_1 + w_{2,5}P_2 + w_{3,5}P_3 + w_{4,5}P_4 + w_{5,5}P_5 + w_{12,5}\varphi[P_1, P_2] + w_{13,5}\varphi[P_1, P_3]}{|w_{1,5}| + |w_{2,5}| + |w_{3,5}| + |w_{4,5}| + |w_{5,5}| + |w_{12,5}| + |w_{13,5}|}$$

Definition 5.7 (*Structural Influence Function*) transforms multiple interacting influencing agents' preferences or choices to the influenced agent's preference or choice. If this influence function is expressed in a general multiperiod form (taking into account the iteration of mutual influence), then:

$$P_{(i)}(t+1) = \frac{\sum_{j\in\mathbb{N}} w_{(j,i)}P_{(j)}(t) + \sum_{\mathfrak{s}\in\mathbb{S}_{[\mathbb{N}](i)}(t)} w_{(\mathfrak{s},i)}\varphi[\mathfrak{s}]}{\sum_{j\in\mathbb{N}} |w_{(j,i)}| + \sum_{\mathfrak{s}\in\mathbb{S}_{[\mathbb{N}](i)}(t)} |w_{(\mathfrak{s},i)}|} \quad i \in \mathbb{N}$$

In which $\mathbb{S}_{[\mathbb{N}](i)}(t)$ represents the set of all structures (influencing relationships) among influencing agents in the set of all agents $\mathbb{N}$ from the perspective of agent i at tth mutual influence, $\varphi[\mathfrak{s}]$ ($\mathfrak{s} \in \mathbb{S}_{[\mathbb{N}](i)}(t)$) represents the *structural influence* produced by influencing relationship $\mathfrak{s}$, and $w_{(\mathfrak{s},i)}$ represents the weight of influence from influencing relationship $\mathfrak{s}$ to agent i.

The above is just a general mathematical expression of how to address *structural influence*, a specific *structural influence function* can be found in the next chapter.

5.4   The Relationships Among the Three Levels of Influence

We have to admit that this book mainly provides a mathematical modelling and graphical expression for *individual, coalitional and structural influence*, and we have not evaluated and compared the relative merits of the above different influence models (functions). A critical reason is that the relationships among the *three levels of influence* are not substitutional or competitive, but all of them together can fully describe the complex features of influence in real-world settings.[12]

---

[11] Both of which belong to the set of influencing agents from the perspective of the influenced agent whose updated preference or choice we are trying to acquire. Although they are all influencing agents from the perspective of the influenced agent, they can still be classified while comparing them to each other if there are also influencing relationships between them.

[12] All of them are components of the real-world influence.

How to Address the Interplay of Individual, Coalitional and Structural Influence: A Probability-Based Approach

6

In reality, the *three levels of influence* can work together. To better understand the mechanisms of the *three levels of influence*, and in particular their mixed effects, we build mathematical models using a probability-based (probabilistic) choice approach and provide an example of group decision-making with multiple sources of influence to illustrate how to address *individual, coalitional and structural influence* using the models we build.

Specifically, we use the probabilities of choosing different alternatives (candidates) of an agent to represent his or her decision-making preference, which is a nonordering and cardinal approach.

Example 6.1 (The Mixed Effects of Individual, Coalitional and Structural Influences) As shown in Fig. 6.1, we assume a group decision-making system with eight agents (No. 1–8) making a choice on an issue with three alternatives: $\{a, b, c\}$. From the perspective of the agent (No. 8) at the bottom, he or she is simultaneously influenced by eight agents, with four agents (No. 1, 2, 4, and 6) choosing (supporting) a, one agent (No. 3) choosing (supporting) b, and three agents (No. 5, 7, and 8 himself or herself[1]) choosing (supporting) c. What's more, the agent (No. 8) being influenced observes or believes that there are two influencing relationships among the three influencing agents (No. 1, 4, and 6) choosing a and one influencing relationship between the two influencing agents (No. 5 and 7) choosing c. Assume that the influencing relationships are specific as "following" (i.e. the influenced one follows what the influencer says or does[2]); and simply assume that all the weights of influence are positive.

---

[1] Originally or before the influence.

[2] The influenced one does what the influencer does and says what the influencer says.

© The Author(s), under exclusive license to Springer Nature Switzerland AG 2026
H. Luo, *Influence Models in Group Decision-Making*, Synthesis Lectures on
Computer Science, https://doi.org/10.1007/978-3-032-01352-1_6

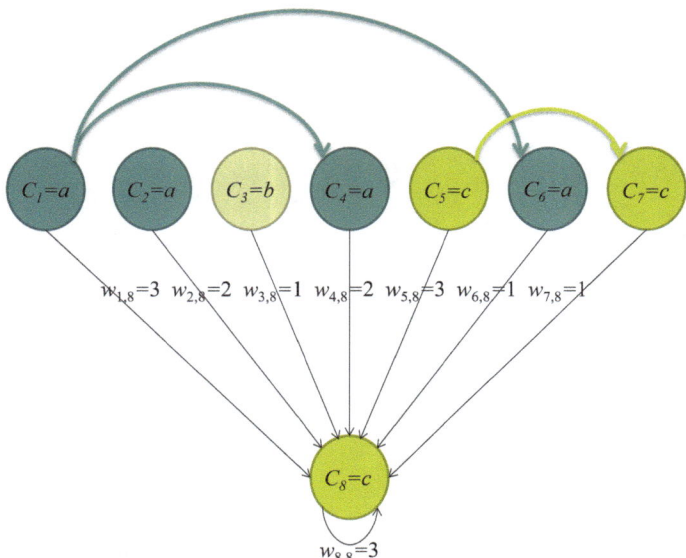

Fig. 6.1 An example of the mixed effects of individual, coalitional and structural influence. *Notes* C represents the choice of each agent. The weight of each agent's influence is marked on the link representing that influence

6.1   How to Address Individual Influences

If just considering the independent influences from all influencing agents but not the coalitions and structures[3] among them, it is easy to form a linear weighted average function to obtain the influence result.

Definition 6.1 (*Probability-based Individual Influence Function*) Let $P_{o(i)}(t+1)$ be the preference of agent i for alternative o after tth mutual influence, which can be expressed by a probability of agent i choosing alternative o. It will be influenced by all agents choosing alternative o at tth mutual influence, and according to their weights of influence on agent i (mathematically, the weighted ratio of influencing agents choosing alternative o):

$$P_{o(i)}(t+1) = \frac{\sum_{C_{(j)}(t)=o} w_{(j,i)}}{\sum_{j\in\mathbb{N}} w_{(j,i)}} \quad i \in \mathbb{N}, o \in \mathbb{M}$$

In which $C_{(j)}(t) = o$ means the choice of agent j at tth mutual influence is alternative o, $\mathbb{N}$ is the set of all agents and $\mathbb{M}$ is the set of all alternatives.

---

[3] i.e. influencing relationships.

In the Example 6.1, assume that $P_{a(8)}$, $P_{b(8)}$, $P_{c(8)}$ are respectively the probabilities of the influenced agent 8 choosing alternatives a, b, c after the mutual influence:

$$
\begin{aligned}
P_{a(8)} &= \frac{w_{1,8} + w_{2,8} + w_{4,8} + w_{6,8}}{w_{1,8} + w_{2,8} + w_{3,8} + w_{4,8} + w_{5,8} + w_{6,8} + w_{7,8} + w_{8,8}} \\
&= \frac{3 + 2 + 2 + 1}{3 + 2 + 1 + 2 + 3 + 1 + 1 + 3} \\
&= \frac{8}{16} \\
&= 0.5000
\end{aligned}
$$

$$
\begin{aligned}
P_{b(8)} &= \frac{w_{3,8}}{w_{1,8} + w_{2,8} + w_{3,8} + w_{4,8} + w_{5,8} + w_{6,8} + w_{7,8} + w_{8,8}} \\
&= \frac{1}{3 + 2 + 1 + 2 + 3 + 1 + 1 + 3} \\
&= \frac{1}{16} \\
&= 0.0625
\end{aligned}
$$

$$
\begin{aligned}
P_{c(8)} &= \frac{w_{5,8} + w_{7,8} + w_{8,8}}{w_{1,8} + w_{2,8} + w_{3,8} + w_{4,8} + w_{5,8} + w_{6,8} + w_{7,8} + w_{8,8}} \\
&= \frac{3 + 1 + 3}{3 + 2 + 1 + 2 + 3 + 1 + 1 + 3} \\
&= \frac{7}{16} \\
&= 0.4375
\end{aligned}
$$

6.2 How to Address Structural Influences

If considering the influencing effects from structures (i.e. influencing relationships) among influencing agents, there are still different perspectives to understand and address this kind of influence, because the human mind is inherently complicated. Different people have various personalities and value systems. Even for a single person, his or her cognition will be different under changing environments, spaces, times, emotions, states of health,[4] and other variables. Here, we provide a simple solution (analytic framework) with two perspectives (angles of view) to address *structural influence*: one is weakening the weights of influence from the "followers" within influencing agents, the other is intensifying the weights of influence from the "leaders" within influencing agents.

From one angle of view, the influenced agent may think that some influencing agents also follow some other influencing agents and do not have independent ideas or a mind

---

[4] Even good sleep or bad sleep matters.

of their own (thus, looking down on them as just "followers"); therefore, after perceiving such influencing relationships among the influencing agents or identifying such influencing agents, the influenced agent may be inclined to "ignore" or "belittle" them, more specifically (especially in mathematics), to decrease the weights of influence from the "influenced" influencing agents (to a certain degree).

From the other angle of view, the influenced agent may focus on the influencers or uninfluenced ones but not the influenced ones[5] within influencing agents, thinking that the reason why some influencing agents always insist on their (original) ideas and even can influence other (influencing) agents is that they indeed have correct or better beliefs (like truth-holders) or that they are very influential, forceful or powerful (like opinion leaders or authorities[6]), and believe that it is safer or more advantageous to adopt or refer to the preferences[7] or choices[8] of these "pure" influencing agents (like wise men or leaders); therefore, after perceiving such influencing relationships among the influencing agents or identifying such influencing agents, the influenced agent may be inclined to "emphasize" them, more specifically (especially in mathematics), to increase the weights of influence from the "uninfluenced" influencing agents (to a certain degree).[9]

To realize this solution for *structural influence* to obtain the influence result, we should first distinguish between the influencing agents with uninfluenced (original) preferences and those with influenced (updated) preferences[10] (that is, they are also influenced by other influencers). In order to achieve such a classification, some more nuanced variables should be defined and added to the probability-based influence function.

Definition 6.2 (*Probability-based Structural Influence Function*) Let α, β be a pair of *structural influence coefficients*, which are timed respectively by the weights of "uninfluenced" influencing agents and the weights of "influenced" influencing agents. The *structural influence coefficients* should satisfy $\alpha + \beta = 1$, and by common sense $\alpha \geq \beta$. Let $P^{\mathbb{S}}_{o(i)}(t+1)$ be the probability of agent i choosing alternative o after tth mutual influence considering $\mathbb{S}$*tructural influence*, which will be influenced by all agents choosing alternative o at tth mutual influence according to their weights of influence on agent i, and also affected by these agents' "roles" in the influencing relationships among the influencing agents (from the perspective of agent i):

---

[5] In another word, focusing on the "influencing" influencing agents but not the "influenced" influencing agents.

[6] According to Max Weber, there are three types of legitimate authority: traditional, rational-legal, and charismatic.

[7] Such as beliefs, opinions.

[8] Or other kinds of behaviors.

[9] It is common for people to emphasize and respect "leaders" more than "followers".

[10] i.e. with their own original preferences being influenced.

$$P^{\mathbb{S}}_{o(i)}(t+1) = \frac{\alpha \sum_{C_{(j)}(t)=o=C_{(j)}(t-1)} w_{(j,i)} + \beta \sum_{C_{(j)}(t)=o\neq C_{(j)}(t-1)} w_{(j,i)}}{\alpha \sum_{C_{(j)}(t)=C_{(j)}(t-1)} w_{(j,i)} + \beta \sum_{C_{(j)}(t)\neq C_{(j)}(t-1)} w_{(j,i)}} \quad i \in \mathbb{N}, o \in \mathbb{M}$$

In which $C_{(j)}(t) = C_{(j)}(t-1)$ means the choice of agent j after $t-1$th mutual influence does not change, $C_{(j)}(t) \neq C_{(j)}(t-1)$ means the choice of agent j after $t-1$th mutual influence changes.

Usually, whose influenced preference or choice we are trying to acquire, then whose perspective (how this agent perceives the influencing relationships among others) should be adopted in the processing of influence.

In the Example 6.1, assume that $P^{\mathbb{S}}_{a(8)}, P^{\mathbb{S}}_{b(8)}, P^{\mathbb{S}}_{c(8)}$ are respectively the probabilities of the influenced agent 8 choosing alternatives a, b, c after the mutual influence considering $\mathbb{S}$tructural influence and assume three pairs of *structural influence coefficients*:

$$P^{\mathbb{S}}_{a(8)} = \frac{\alpha w_{1,8} + \alpha w_{2,8} + \beta w_{4,8} + \beta w_{6,8}}{\alpha w_{1,8} + \alpha w_{2,8} + \alpha w_{3,8} + \beta w_{4,8} + \alpha w_{5,8} + \beta w_{6,8} + \beta w_{7,8} + \alpha w_{8,8}}$$

$$= \begin{cases} \frac{0.5\times3+0.5\times2+0.5\times2+0.5\times1}{0.5\times3+0.5\times2+0.5\times1+0.5\times2+0.5\times3+0.5\times1+0.5\times1+0.5\times3} = 0.5000 &, \alpha = 0.5, \beta = 0.5 \\ \frac{0.8\times3+0.8\times2+0.2\times2+0.2\times1}{0.8\times3+0.8\times2+0.8\times1+0.2\times2+0.8\times3+0.2\times1+0.2\times1+0.8\times3} = 0.4423 &, \alpha = 0.8, \beta = 0.2 \\ \frac{1\times3+1\times2+0\times2+0\times1}{1\times3+1\times2+1\times1+0\times2+1\times3+0\times1+0\times1+1\times3} = 0.4167 &, \alpha = 1, \beta = 0 \end{cases}$$

$$P^{\mathbb{S}}_{b(8)} = \frac{\alpha w_{3,8}}{\alpha w_{1,8} + \alpha w_{2,8} + \alpha w_{3,8} + \beta w_{4,8} + \alpha w_{5,8} + \beta w_{6,8} + \beta w_{7,8} + \alpha w_{8,8}}$$

$$= \begin{cases} \frac{0.5\times1}{0.5\times3+0.5\times2+0.5\times1+0.5\times2+0.5\times3+0.5\times1+0.5\times1+0.5\times3} = 0.0625 &, \alpha = 0.5, \beta = 0.5 \\ \frac{0.8\times1}{0.8\times3+0.8\times2+0.8\times1+0.2\times2+0.8\times3+0.2\times1+0.2\times1+0.8\times3} = 0.0769 &, \alpha = 0.8, \beta = 0.2 \\ \frac{1\times1}{1\times3+1\times2+1\times1+0\times2+1\times3+0\times1+0\times1+1\times3} = 0.0833 &, \alpha = 1, \beta = 0 \end{cases}$$

$$P^{\mathbb{S}}_{c(8)} = \frac{\alpha w_{5,8} + \beta w_{7,8} + \alpha w_{8,8}}{\alpha w_{1,8} + \alpha w_{2,8} + \alpha w_{3,8} + \beta w_{4,8} + \alpha w_{5,8} + \beta w_{6,8} + \beta w_{7,8} + \alpha w_{8,8}}$$

$$= \begin{cases} \frac{0.5\times3+0.5\times1+0.5\times3}{0.5\times3+0.5\times2+0.5\times1+0.5\times2+0.5\times3+0.5\times1+0.5\times1+0.5\times3} = 0.4375 &, \alpha = 0.5, \beta = 0.5 \\ \frac{0.8\times3+0.2\times1+0.8\times3}{0.8\times3+0.8\times2+0.8\times1+0.2\times2+0.8\times3+0.2\times1+0.2\times1+0.8\times3} = 0.4808 &, \alpha = 0.8, \beta = 0.2 \\ \frac{1\times3+0\times1+1\times3}{1\times3+1\times2+1\times1+0\times2+1\times3+0\times1+0\times1+1\times3} = 0.5000 &, \alpha = 1, \beta = 0 \end{cases}$$

When $\alpha = 0.5, \beta = 0.5$, *structure influence* is actually not considered (in some sense); when $\alpha = 1, \beta = 0$, the weights of influence from the "followers" are totally eliminated, which is equivalent to (completely) ignoring the "followers", which is an extreme assumption or setting; and we also assume a relatively mild case $\alpha = 0.8, \beta = 0.2$. We can find that the probability of the influenced agent 8 choosing alternative a is reduced after considering *structural influence*, which makes sense as agent 8 finds or believes that half of the influencing agents choosing (supporting) alternative a are just "followers", without independent ideas or a mind of their own; both the probabilities of the influenced agent 8 choosing alternatives b and c increase after considering *structural influence*, which makes sense as agent 8 finds or believes that no influencing agent choosing (supporting) alternative b is a "follower" and only one influencing agent choosing (supporting) alternative c is a "follower".

6.3   How to Address Coalitional Influences

If the influencing effects from coalitions (coalitional agents) among influencing agents are also considered, then the mixed effects with the influencing effects from structures (structured agents) should be addressed to obtain the compound influence result.

Definition 6.3 (*Probability-based Coalitional Influence Function*) Let $P^{\mathbb{SC}}_{o(i)}(t)$ be the probability of agent i choosing alternative o at tth mutual influence considering both *$\mathbb{S}$tructural influence* and *$\mathbb{C}$oalitional influence*. $P^{\mathbb{S}}_{o(i)}(t)$ is the probability of agent i choosing alternative o at tth mutual influence after considering *$\mathbb{S}$tructural influence* but before considering *coalitional influence*.

If *coalitional influence* works in a way similar to majority rule, which means that once the probability of agent i choosing alternative o (i.e. the weighted ratio of influencing agents choosing alternative o) exceeds $\frac{1}{2}$, then the resulting choice for agent i will be alternative o for sure (which means that the probability of agent i choosing alternative o enlarges to 100%, while the probability of agent i choosing any other alternative should be reduced to 0, to ensure that the sum of the probabilities is 100%); otherwise, unsure, but depending on the comparison of the probabilities of agent i choosing different alternatives (which essentially depends on the comparison of the weighted ratios of influencing agents choosing different alternatives from the perspective of agent i).

$$
P^{\mathbb{SC}}_{o(i)}(t) = \begin{cases} 1 & , P^{\mathbb{S}}_{o(i)}(t) > \frac{1}{2} \\ P^{\mathbb{S}}_{o(i)}(t) & , \bigwedge_{o' \in \mathbb{M}} P^{\mathbb{S}}_{o'(i)}(t) \leq \frac{1}{2} \quad i \in \mathbb{N}, o \in \mathbb{M} \\ 0 & , otherwise \end{cases}
$$

As shown in Fig. 6.2, the two curves of different colors mean that if $P^{\mathbb{S}}_{o(i)}(t) \leq \frac{1}{2}$, there are two different possible outcomes for $P^{\mathbb{SC}}_{o(i)}(t)$, depending on whether there is another alternative whose probability of being chosen reaches the majority, i.e. $> \frac{1}{2}$; while if $P^{\mathbb{S}}_{o(i)}(t) > \frac{1}{2}$, reaching the majority, then $P^{\mathbb{SC}}_{o(i)}(t)$ will be 100% uniformly.

If *coalitional influence* works in a way similar to plurality rule, which means that once the probability of agent i choosing alternative o (i.e. the weighted ratio of influencing agents choosing alternative o) exceeds $\frac{1}{2}$, then alternative o will definitely (100%) be the resulting choice for agent i (as a majority is certainly a plurality); and once the probability of agent i choosing alternative o falls short of $\frac{1}{m}$ (assume there are m alternatives), then alternative o will definitely not be the resulting choice for agent i (as less than $\frac{1}{m}$, i.e. the average probability, is certainly not a plurality); otherwise, uncertain, but depending on the comparison of the probabilities of agent i choosing different alternatives (which essentially depends on the comparison of the weighted ratios of influencing agents choosing different alternatives from the perspective of agent i).

Fig. 6.2 Decision-making
(voting) probability before and
after coalitional influence
referring to majority rule

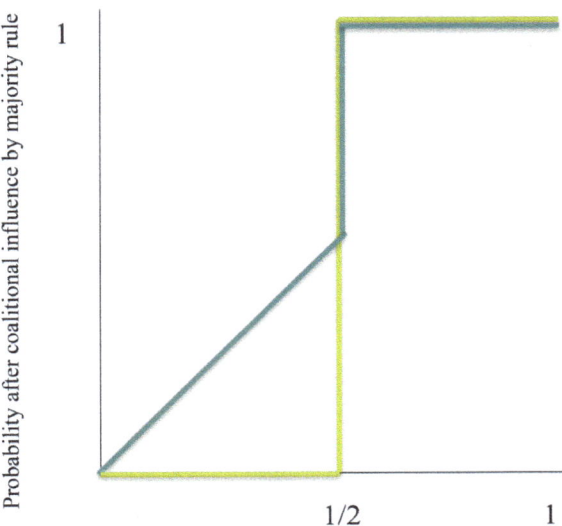

Probability before coalitional influence

$$
P_{o(i)}^{SC}(t) =
\begin{cases}
1 & , P_{o(i)}^{S}(t) > \frac{1}{2} \\[4pt]
1 & , \frac{1}{2} \geq P_{o(i)}^{S}(t) > \frac{1}{m} \wedge [\bigwedge_{o' \in \mathbb{M}\setminus\{o\}} P_{o(i)}^{S}(t) > P_{o'(i)}^{S}(t)] \\[4pt]
0 & , \frac{1}{2} \geq P_{o(i)}^{S}(t) > \frac{1}{m} \wedge [\bigvee_{o' \in \mathbb{M}\setminus\{o\}} P_{o(i)}^{S}(t) < P_{o'(i)}^{S}(t)] \\[4pt]
0 & , otherwise
\end{cases}
\quad i \in \mathbb{N}, o \in \mathbb{M}
$$

As shown in Fig. 6.3, the two curves of different colors mean that when $\frac{1}{2} \geq P_{o(i)}^{S}(t) \geq \frac{1}{m}$, there are two different possible outcomes for $P_{o(i)}^{SC}(t)$, depending on whether the probability of choosing this alternative is the highest among the probabilities of all the alternatives being chosen; when $P_{o(i)}^{S}(t) > \frac{1}{2}$, $P_{o(i)}^{SC}(t)$ will be 100% uniformly; when $P_{o(i)}^{S}(t) < \frac{1}{m}$, $P_{o(i)}^{SC}(t)$ will be 0% uniformly.

It should be noted that there is a small-probability special case for the effect of *coalitional influence* working in a way similar to plurality rule: if the probabilities of several alternatives being chosen tie for highest, then they equally share 100% probability; if there are $\dot{m}$ tied alternatives, each of their probabilities to be chosen is $\frac{1}{\dot{m}}$.

In the Example 6.1, when the *structural influence coefficients* $\alpha = 0.5$, $\beta = 0.5$, we have the preference of the influenced agent 8 after *structural influence*[11] expressed as a probability distribution: $P_{a(8)}^{S} = 0.5000$, $P_{b(8)}^{S} = 0.0625$, $P_{c(8)}^{S} = 0.4375$. If *coalitional influ-*

---

[11] As mentioned earlier, when $\alpha = \beta = 0.5$, it is actually equivalent to not considering *structural influence*.

Fig. 6.3 Decision-making
(voting) probability before and
after coalitional influence
referring to plurality rule

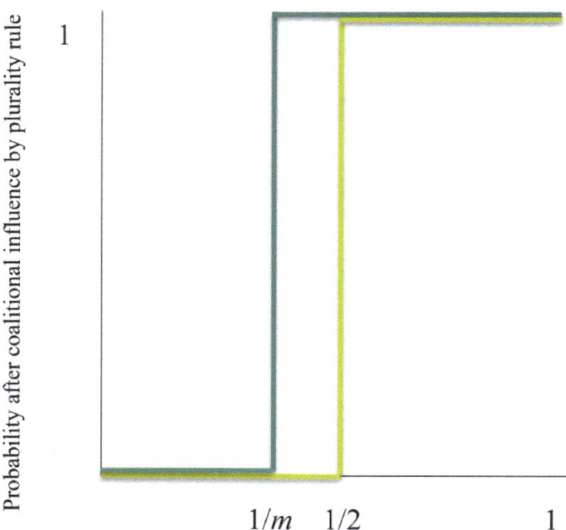

Probability before coalitional influence

ence works in a way similar to majority rule, then $P_{a(8)}^{SC} = P_{a(8)}^{S} = 0.5000$, $P_{b(8)}^{SC} = P_{b(8)}^{S} =$
0.0625, $P_{c(8)}^{SC} = P_{c(8)}^{S} = 0.4375$, as no one reaches the majority (i.e. probability $> \frac{1}{2}$), despite
it is very close for alternative a;[12] while if *coalitional influence* works in a way similar to
plurality rule, then $P_{a(8)}^{SC} = 1$, $P_{b(8)}^{SC} = 0$, $P_{c(8)}^{SC} = 0$, as $P_{a(8)}^{S}$ is larger than both $P_{b(8)}^{S}$ and
$P_{c(8)}^{S}$, becoming the plurality.

When the *structural influence coefficients* $\alpha = 0.8$, $\beta = 0.2$, we have the preference
of the influenced agent 8 after *structural influence* expressed as a probability distribu-
tion: $P_{a(8)}^{S} = 0.4423$, $P_{b(8)}^{S} = 0.0769$, $P_{c(8)}^{S} = 0.4808$. If *coalitional influence* works in a
way similar to majority rule, then $P_{a(8)}^{SC} = P_{a(8)}^{S} = 0.4423$, $P_{b(8)}^{SC} = P_{b(8)}^{S} = 0.0769$, $P_{c(8)}^{SC} =$
$P_{c(8)}^{S} = 0.4808$, as no one reaches the majority (i.e. probability $> \frac{1}{2}$); while if *coalitional
influence* works in a way similar to plurality rule, then $P_{a(8)}^{SC} = 0$, $P_{b(8)}^{SC} = 0$, $P_{c(8)}^{SC} = 1$, as
$P_{c(8)}^{S}$ is larger than both $P_{a(8)}^{S}$ and $P_{b(8)}^{S}$, becoming the plurality.

When the *structural influence coefficients* $\alpha = 1$, $\beta = 0$, we have the preference of the
influenced agent 8 after *structural influence* expressed as a probability distribution: $P_{a(8)}^{S} =$
0.4167, $P_{b(8)}^{S} = 0.0833$, $P_{c(8)}^{S} = 0.5000$. If *coalitional influence* works in a way simi-
lar to majority rule, then $P_{a(8)}^{SC} = P_{a(8)}^{S} = 0.4167$, $P_{b(8)}^{SC} = P_{b(8)}^{S} = 0.0833$, $P_{c(8)}^{SC} = P_{c(8)}^{S} =$
0.5000, as no one reaches the majority (i.e. probability $> \frac{1}{2}$), although alternative c is so
close; while if *coalitional influence* works in a way similar to plurality rule, then $P_{a(8)}^{SC} =$
0, $P_{b(8)}^{SC} = 0$, $P_{c(8)}^{SC} = 1$, as $P_{c(8)}^{S}$ is larger than both $P_{a(8)}^{S}$ and $P_{b(8)}^{S}$, becoming the plurality.

---

[12] 0.5001 will be enough.

Part III

Influence Across Agents and Issues in Combinatorial and Collective Decision-Making

Influence from multiple agents while making decisions on multiple issues.

Multiple Sources of Influence Across Agents and Issues

Previous studies have mainly discussed the influence on an agent of multiple agents while making decisions on a single issue (usually, in the context of social networks) [11, 14, 17–24, 37, 55, 58] or the dependency on[1] multiple issues of an issue decided by a single agent [3, 4, 26, 56, 62, 63] (typically, in the framework of CP-nets).[2] However, in reality, an agent's preference/choice on an issue can be simultaneously influenced by multiple agents' preferences/choices on multiple issues.[3] Thus, the source (origin) of influence is a more general entity, involving both agents and issues. However, previous studies have less discussed how to address the influence from more than one source across both multiple agents and multiple issues. Therefore, we propose a framework of combinatorial and collective decision-making with influence across agents and issues illustrated by Example 7.1, as follows:

---

[1] Which can also be understood as "influence by".

[2] A few studies (such as [48, 49]) described influence in a multi-agent and multi-issue setting, but just discussed influence among agents and dependency among issues separately, not considering the influence across both agents and issues.

[3] For a simple example, agent i's preference/choice on issue Z may be influenced by agent j's preference/choice on a former issue X and agent k's preference/choice on another former issue Y.

© The Author(s), under exclusive license to Springer Nature Switzerland AG 2026
H. Luo, *Influence Models in Group Decision-Making*, Synthesis Lectures on
Computer Science, https://doi.org/10.1007/978-3-032-01352-1_7

Example 7.1 (A General Example of Influences across Agents and Issues) Assume a case of multi-agent and multi-issue decision-making with a set of agents $\{1, 2, 3\}$ making decisions on a set of issues $\{X, Y, Z\}$, each with three alternatives, as shown in Fig. 7.1.

While agent 3 is making a decision on issue Z, it is possible that his or her preference/choice will be influenced by agent 2's preference/choice on the same issue Z (which is an influence between agents making decisions on a single issue) and also influenced by his or her own preference/choice on former issue Y (which is a dependency between issues decided by a single agent), while at the same time, influenced by agent 2's preference/choice on former issue Y and also influenced by agent 1's preference/choice on former issue X (which are influences across both agents and issues).

When an agent's preference/choice on an issue is simultaneously influenced by more than one agent's preference/choice on more than one issue, especially when different sources of influence have opposite influencing directions (positive or negative) and varied influencing strengths (stronger or weaker), how to determine the result of the multiple influences is an important, yet not simple, question. It is relatively easy to set the rule of influence from multiple agents to another agent while making a decision on a single issue[4] or the rule of dependency on multiple (former) issues of another (later) issue decided by a single agent,[5] but it is much more complicated to design a rule of influence from more than one source across both multiple agents and multiple issues.

In fact, the influence from multiple sources across both agents and issues is very common in real-world situations. It is oversimplifying to assume that each agent's preference/choice on each issue is influenced by other agents' preferences/choices only on the same issue (that is, assume that there are only influences in the horizontal dimension, as shown in Fig. 7.1), or influenced only by his or her own preferences/choices on other former issues (that is, assume that there are only influences in the vertical dimension, as shown in Fig. 7.1), or only influenced by another agent's preference/choice on another issue (that is, assume that there is only one-to-one influence in the diagonal dimension rather multiple influences) at a time (or in a round).

7.1   Stronger and Weaker Influences Across Agents and Issues

The reality is that the preference/choice of a person on an issue can be influenced by the preferences/choices of different people on different issues to different degrees (corresponding to stronger or weaker influences) at the same time. Likewise, the reasoning[6] and answer of an artificial intelligence on a question can be influenced (e.g. in the form of learning) by

---

[4] Of course, it is more simple to set the rule of influence from one agent to another agent while making a decision on an issue.

[5] Of course, it is more simple to set the rule of dependency on one issue of another issue decided by an agent.

[6] Knowledge representation underlying.

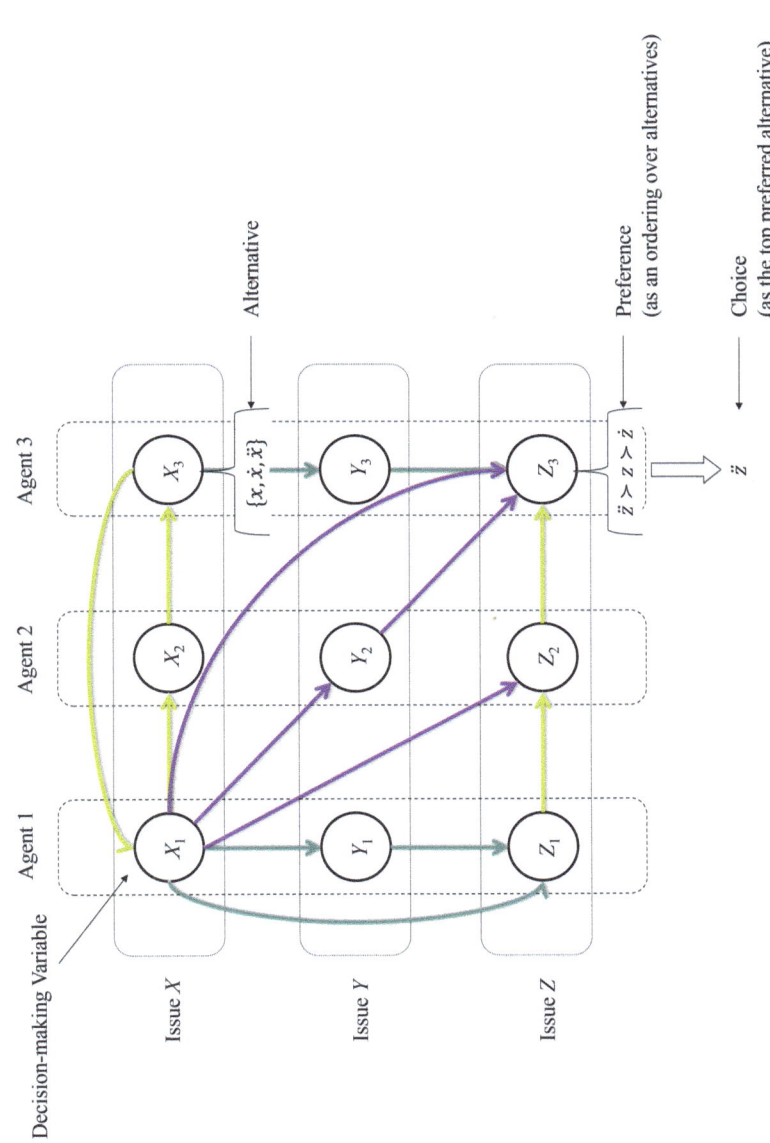

Fig. 7.1 Influences across multiple agents and multiple issues in combinatorial and collective decision-making. *Notes* Each node represents a decision-making variable (i.e. the preference/choice of an agent while making a decision on an issue). Green links represent the dependencies (influences) among multiple issues decided by a single agent; yellow links represent the influences among multiple agents making decisions on a single issue; purple links represent the influences across both multiple agents and multiple issues

the reasoning and answers of different artificial intelligences (such as DeepSeek, ChatGPT, Gemini) on different questions to different extents.

Example 7.2 (A Specific Example of Stronger and Weaker Influences across Agents and Issues: a Family Buying a Car) Suppose a family is choosing a car, and it is a "democratic" rather than a "dictatorial" family in which each family member has a vote (voice). Buying a car is a typical multi-issue decision, because multiple features of the car must be decided, including type (e.g. sedan, SUV, and sports car), manufacturer (e.g. BMW®, Mercedes-Benz®, and Jeep®), color (e.g. black, white, red), etc. Usually, it is full of mutual influence among the preferences or choices of family members on these features (issues).[7] It may be the case that the husband's preference or choice for the type of the car is strongly influenced by his wife's preference or choice of car manufacturer: if his wife wants a BMW®, he will be much more inclined to buy a sedan than an SUV, but if his wife wants a Jeep®, he will be much more inclined to buy an SUV than a sedan; meanwhile, his preference or choice for the type of the car is slightly influenced by his children' preference or choice of car color: if his children want a black car, he will be more inclined to buy a sedan than a sports car, but if his children want a red car, he will be more inclined to buy a sports car than a sedan.

7.2   Positive and Negative Influences Across Agents and Issues

Moreover, a person's preference/choice on an issue can be simultaneously positively influenced by the preferences/choices of some people (such as friends and family) and negatively influenced by the preferences/choices of some other people (such as enemies and opponents) on different issues at the same time.

Example 7.3 (A Specific Example of Positive and Negative Influences across Agents and Issues: the UN Security Council Voting) Each member state in the UN Security Council has full motivation to persuade and influence the votes of other member states in order to achieve desired voting results and maximize its own state interests. During the process of the UN Security Council voting, there are both positive influences and negative influences among member states[8] due to uniform (similar) or conflicting state interests and the existence of

---

[7] For instance, the husband's preference or choice may be influenced by his wife's preference or choice, and the wife's preference may be affected by her children's preferences. Although all family members usually have their own original preferences about the features of the car, during the interaction with family members, their preferences may be influenced by others, being empathetic with each other.

[8] The decision-making process of the UN Security Council involves various influences among its member states, including both positive influences among allies and negative influences among opponents.

confronting alliances (camps).[9] Member states in the same camp usually support each other, in another word, positively influencing each other. For instance, the United Kingdom usually casts the same vote as the United States. However, member states from confronting camps usually oppose each other, in another word, negatively influencing each other. For instance, the former Soviet Union[10] and present-day Russia usually veto the draft resolutions proposed by the United States and other NATO members.

The UN Security Council voting is also a typical multi-issue decision, with each draft resolution corresponding to one issue. Usually, the vote of a member state on a draft resolution will be positively influenced by the votes (or opinions) of its allies on the same draft resolution and the votes of its allies and itself on former relevant draft resolutions (with the same or similar subjects) and negatively influenced by the votes (or opinions) of its opponents on the same draft resolution and the votes of its opponents on former relevant draft resolutions.

7.3   Three Rules for Addressing Multiple Influences Across Agents and Issues

In a multi-agent and multi-issue decision-making context, each influencing and influenced entity (represented by a node as shown in Fig. 7.1), which can be defined as a decision-making variable[11] (typically as preference or choice), needs two coordinates (one is the issue-coordinate and the other is the agent-coordinate) to be located (as shown in Fig. 7.1), i.e. to know which agent is making a decision and this agent is making a decision on which issue. We define the decision-making variable and some relevant (background) variables in a combinatorial and collective decision-making context ("combinatorial" means there are multiple issues with dependencies among them, and "collective" means there are multiple agents making decisions) as follows:

Definition 7.1 (*Combinatorial and Collective Decision-making Society with Influence Across Agents and Issues*) Assume a society $\mathbb{S} = \{\mathbb{N}, \mathbb{I}, \mathbb{P}, \mathbb{C}, \mathbb{W}\}$, where $\mathbb{N} = \{1, 2, ..., n\}$ is the set of all agents (a general term that can represent a person or an artificial intelligence in nature and that can represent a decision-maker, a voter, a game player, etc. in function); $\mathbb{I} = \{I(1), I(2), ..., I(m)\}$ is the set of all issues (or all features of a multi-feature decision,

---

[9] Such as the confrontation between the NATO led by the United States and the Warsaw Pact led by the former Soviet Union during the Cold War, the antagonism between the Western world and Russia nowadays.

[10] During the first 10 years of the United Nations, the former Soviet Union representative, Andrei Gromyko, even had the nickname "Mr. No". This is understandable because, during that period, the socialist camp (bloc) was the absolute minority in the United Nations and the former Soviet Union was mainly in a defensive position.

[11] Each represents an agent making a decision on an issue.

each issue or feature has several alternatives to choose from, and the sets of alternatives of different issues or features are usually different); then, there are $n \times$ m decision-making variables (n agents times with m issues) in total; $\mathbb{P} = \{P_{(1)}(1), P_{(1)}(2), ..., P_{(1)}(\mathrm{m}), P_{(2)}(1), P_{(2)}(2), ..., P_{(2)}(\mathrm{m}), ..., P_{(n)}(1), P_{(n)}(2), ..., P_{(n)}(\mathrm{m})\}$ is the set of all agents' preferences (such as preference orderings, utilities, beliefs, opinions, decision-making probabilities) on all issues; $\mathbb{C} = \{C_{(1)}(1), C_{(1)}(2), ..., C_{(1)}(\mathrm{m}), C_{(2)}(1), C_{(2)}(2), ..., C_{(2)}(\mathrm{m}), ..., C_{(n)}(1), C_{(n)}(2), ..., C_{(n)}(\mathrm{m})\}$ is the set of all agents' choices on all issues, i.e. counting from agent 1's preference/choice on issue 1 to agent n's preference/choice on issue m, in which $P_{(i)}(q)$ represents the preference of agent i on issue q, $C_{(i)}(q)$ represents the choice of agent i on issue q ($i \in \mathbb{N}, q \in \mathbb{I}$); $\mathbb{W}$ is the matrix whose entries are the weights of influence between each two decision-making variables, $\mathbb{W} = [w_{(i,j)}(q, h)]$ ($i, j \in \mathbb{N}, q, h \in \mathbb{I}$), in which $w_{(i,j)}(q, h)$ means the weight of influence from agent i's preference/choice on issue q to agent j's preference/choice on issue h, the weight value indicates both the strength and polarity of the influence, $w_{(i,j)}(q, h) > 0$ means a positive influence, $w_{(i,j)}(q, h) < 0$ means a negative influence, $w_{(i,j)}(q, h) = 0$ means there is no influence from agent i's preference/choice on issue q to agent j's preference/choice on issue h, and the higher $|w_{(i,j)}(q, h)|$, the stronger the influence from agent i's preference/choice on issue q to agent j's preference/choice on issue h.

In this book, three preliminary rules (analytic frameworks or models), *multiple weighted influences*, *one dominant influence*, and *two opposite influences*, addressing the influence from more than one source across both agents and issues, are constructed from different views and mentalities. We first provide a simple comparison of the three rules when it comes to the influence among multiple agents while making decisions on a single issue (that is a multi-agent but single-issue decision-making), illustrated by Example 7.4, then extend the three rules to a multi-agent and multi-issue decision-making context.

Example 7.4 (Multiple Weighted Influences versus One Dominant Influence versus Two Opposite Influences among Multiple Agents Making Decisions on a Single Issue). Assume a case of multi-agent decision-making with eight agents making decisions on an issue with the set of alternatives $\{a, b, c\}$, as shown in Fig. 7.2. While the agent in the middle is making a decision on this issue, he or she is simultaneously influenced by all other agents possessing various preference orderings with diversified weights of influence, in which some influences are positive (such as from friends and family), some other influences are negative (such as from enemies and opponents), and some influences are stronger than others, and even at 3 times the strength[12] of others. All influences are from multiple agents but toward a single issue's decision-making.

---

[12] Which is expressed by the absolute value of the weight of influence in mathematics.

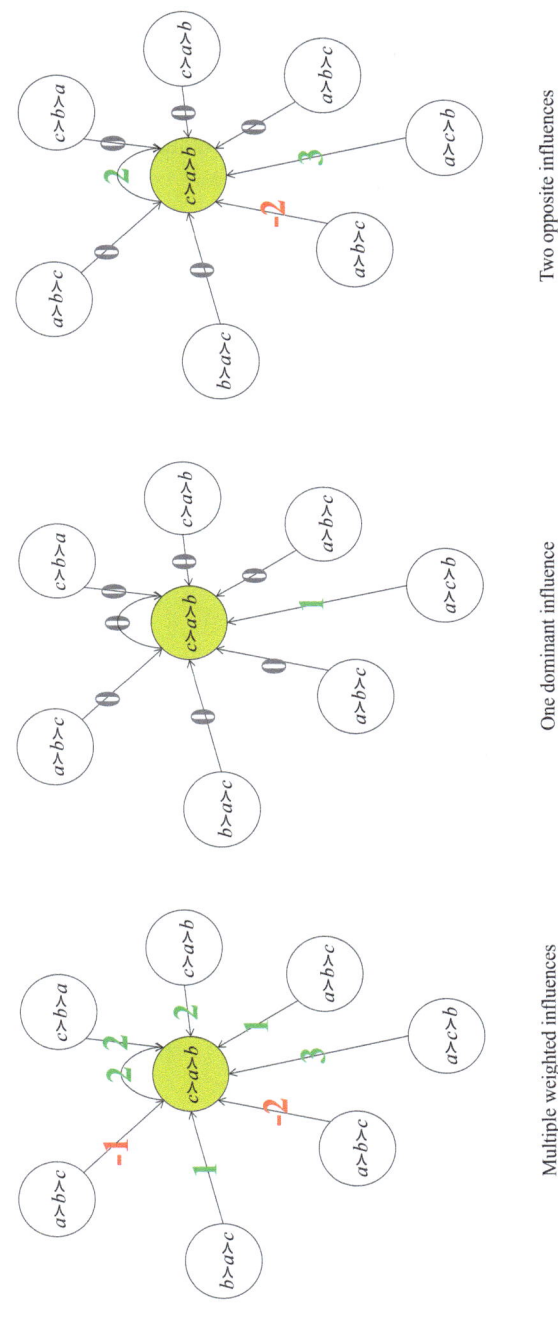

Multiple weighted influences   One dominant influence   Two opposite influences

Fig. 7.2 Multiple weighted influences versus one dominant influence versus two opposite influences among multiple agents making decisions on a single issue. *Notes* The influenced agent is in the middle. The left figure is the real situation of influence and how the influenced agent views the situation of influence from the framework of *multiple weighted influences*, the middle figure is how the influenced agent views the situation of influence from the framework of *one dominant influence*, the right figure is how the influenced agent views the situation of influence from the framework of *two opposite influences*. We consider loops (self influences) here. The weights of positive influence are colored green, and the weights of negative influence are colored red

The comparison of how the three rules address multiple sources of influence across both agents and issues will be discussed in details in the next three chapters, respectively. In short, the rule of *multiple weighted influences* considers each of the multiple influences to different extents and in different directions,[13] without any simplification. In contrast, the rule of *one dominant influence* focuses only on the strongest influence, which has the highest absolute value of weight of influence, but ignores all other influences, reflecting the human mentality of simplifying problems[14] and saving energy. As a compromise or tradeoff (between reducing computational complexity and maintaining information integrity[15]), the rule of *two opposite influences* considers both the strongest positive influence and the strongest negative influence (corresponding to the "closest friend" and "most hated enemy", respectively), and also considers one's own influence; besides, all other influences can be ignored. This rule also makes sense because no matter how we want to simplify problems and save energy, we should at least not ignore the opinions of our best friend, worst enemy and ourself.[16]

---

[13] Positive or negative.

[14] Accordingly, the reasoning and computation can also be simplified.

[15] Or completeness.

[16] Wherever we are, and whenever we are, we should not "forget" ourself.

Multiple Weighted Influences

One rule we design for addressing multiple sources of influence across agents and issues is to assume that the preference/choice of an agent on an issue can be collectively affected by all influencing preferences/choices[1] of more than one agent and on more than one issue, just to different extents and in different directions according to their respective weights of influence.

In fact, the use of the *weight of influence* is a traditional (mainstream) approach to address multiple influences among agents in group decision-making and has been fully practiced in previous works (such as [37, 58]). The empathetic social choice [58] is a cardinal model in which the utility value for each alternative of an influenced agent is the weighted sum of the utilities of all the influencing agents (including his or her "neighbors" and usually himself or herself) for the same alternative.[2] *Social influence functions*, built based on social choice functions (including nonranked social choice methods such as plurality and majority and ranked social choice methods such as Borda count and Condorcet method), can be used to address signed and weighted multiple influences, which means that both positive and negative influences and both stronger and weaker influences (from all influencing agents) are collectively handled to obtain the resulting choice or preference (of an influenced agent). Moreover, the *matrix influence function*, built based on the KSB distance metric [2, 39, 62, 63], can be used to address signed and weighted multiple influences via an ordering-based approach; in order to obtain the resulting preference (of an influenced agent), we define the rule of how to transform each preference ordering (including both the

---

[1] With weight of influence not being equal to 0.

[2] It should be noted that, in the model of empathetic social choice [58], it is assumed that "agents derive utility based on both their own intrinsic preferences and the satisfaction of their neighbors"; when empathetic utilities are formed for agents, their intrinsic utilities remain unchanged, i.e. empathetic utilities and intrinsic utilities are parallel; this treatment differs from our models where when an agent's preference is influenced, his or her preference is updated or the original preference is replaced.

© The Author(s), under exclusive license to Springer Nature Switzerland AG 2026 77
H. Luo, *Influence Models in Group Decision-Making*, Synthesis Lectures on
Computer Science, https://doi.org/10.1007/978-3-032-01352-1_8

preference orderings held by the influencing agents and the preference orderings existing in theory[3]) into a matrix and set a distance metric to compute the distance between any two ordering matrices; then, the feasible (possible) preference ordering that has the smallest weighted sum of distances from all influencing (agents') preference orderings is the resulting preference (of the influenced agent). Since the weights of agents' influences can be positive or negative (typically as friends or enemies) in real-world settings, it will partly play a role in finding the "closest" possible preference from the positively influencing agents' preferences, and partly play a role in finding the "farthest" possible preference from the negatively influencing agents' preferences for the resulting preference.

Traditionally, the empathetic social choice [58], *social influence functions* and *matrix influence function* all discussed the influence from more than one agent making a decision on a single issue but not from more than one source across both multiple agents and multiple issues. However, these models all have potentials to be extended to address multiple sources of influence across both agents and issues in the context of combinatorial and collective decision-making. To achieve this, a precondition is to build a weight matrix whose entries are the weights of influence from each decision-making variable $C_{(i)}(q)$ function environment to each of the other decision-making variables $C_{(j)}(h)$ $(i, j \in \mathbb{N}, q, h \in \mathbb{I})$.

Example 8.1 (A Display of Multiple Weighted Influences across Agents and Issues) Assume a case of multi-agent and multi-issue decision-making with a set of agents $\{1, 2, 3\}$ making decisions on a set of issues $\{X, Y, Z\}$ (which can be the same, similar, or relevant/related issues at different times), each with three alternatives, as shown in Fig. 8.1. Suppose the three agents are now making decisions on issue Z, from the perspective of agent 3, he or she is simultaneously influenced by three agents including agent 1, agent 2, and agent 3 himself or herself making decisions on three issues including issue X, issue Y, and issue Z. Agent 1 is a friend (ally) in the mind of agent 3, thus, agent 1's preferences/choices on current issue Z and on former issue X produce positive influences (with weights of influence respectively as 3 and 1) on agent 3's preference/choice on current issue Z. Agent 2 is an enemy (opponent) in the mind of agent 3, thus, agent 2's preferences/choices on current issue Z and on former issue Y produce negative influences (with weights of influence respectively as -1 and -2) on agent 3's preference/choice on current issue Z. Furthermore, agent 3's preference/choice on current issue Z is influenced by (dependent on) his or her own preferences/choices on former issues X and Y (with weights of influence respectively as 3 and 4).

Usually, a former choice closer to the current time has a higher weight of influence on a current choice than another former choice farther from the current time, because people's memories fade with time[4] (regardless of whether it is the satisfaction of a good past decision

---

[3] If there are m alternatives (candidates), then all feasible preference orderings over them include $m!$ kinds.

[4] Of course, there are exceptions. Some old memories are lasting and profound.

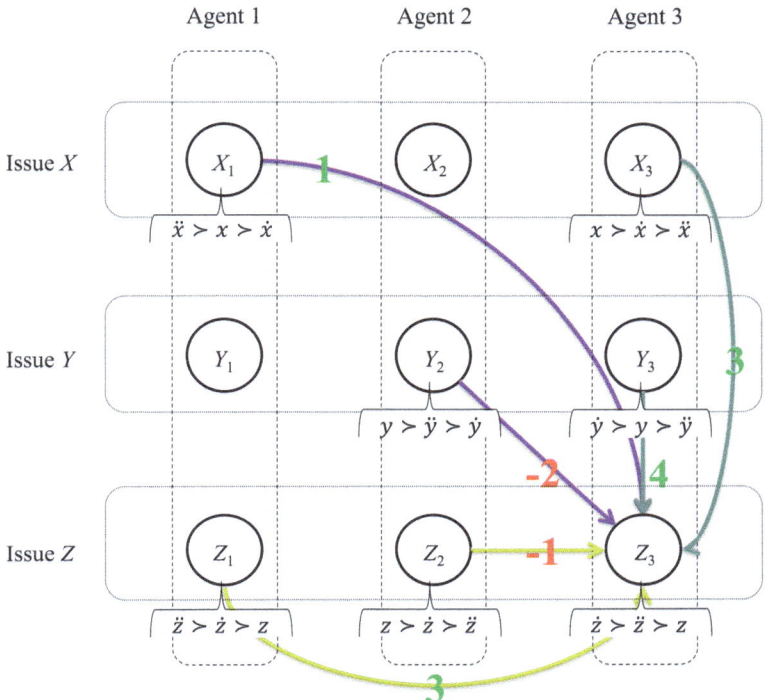

Fig. 8.1 A case of multiple weighted influences across agents and issues. *Notes* Each node represents a decision-making variable (i.e. the preference/choice of an agent while making a decision on an issue). Green links represent the dependencies (influences) among multiple issues decided by a single agent; yellow links represent the influences among multiple agents making decisions on a single issue; purple links represent the influences across both multiple agents and multiple issues. The weights of positive influence are colored green, and the weights of negative influence are colored red. Each influencing preference ordering is displayed (under its corresponding decision-making variable), including the original preference of the influenced agent

or the regret of a bad past decision). According to the rationality hypothesis, people pursue satisfaction and avoid regret. In order to do so, people need to learn from history and use history as a guide, specifically, referring to their own and others' past decisions. As mentioned earlier, it is quite common for people to be influenced not only by other people,[5] around them, but also by themselves. The past self influences the present self, the present self continues to influence the future self[6] and the weight of one's own influence is usually

---

[5] Such as friends and allies, or enemies and opponents.

[6] In another word, the latter self is influenced by the former self.

positive. Only in extreme cases, such as when a person suffers serious setbacks and loses his or her self-confidence completely, could his or her own influence change from strong to weak, and even from positive to negative. Such consideration of one's own influence can explain why some people are difficult to influence by others, while some other people are easy to change, because the former type of people's self influences may have higher weights than the latter.

One Dominant Influence

One more rule we design for addressing multiple sources of influence across agents and issues is to assume that the preference/choice of an agent on an issue will be just influenced by the most predominant influencing preference/choice[1] among all influencing preferences/choices of different agents and on different issues. If the most predominant influence is positive, then the influenced preference/choice becomes the same as the influencing preference/choice; if the most predominant influence is negative, then the influenced preference/choice becomes the opposite of the influencing preference/choice. For example, negative is the opposite of affirmative, $a \succ b \succ c$ is the opposite of $c \succ b \succ a$.

A concept of the *priority of influence* can be proposed to find or determine the most predominant influencing preference/choice. Assume that there are $n \times$ m decision-making variables (corresponding to n agents making decisions on m issues[2]) and that there is a priority of influence from each decision-making variable $C_{(i)}(q)$ function environment to each of the other decision-making variables $C_{(j)}(h)$ ($i, j \in \mathbb{N}$, $q, h \in \mathbb{I}$). The rule of *one dominant influence* assumes that each decision-making variable is influenced (only) by the influencing decision-making variable that has the highest priority (of influence) on it compared to all other influencing decision-making variables.

In fact, a simple and straightforward way to obtain or determine the priority of an influence is to relate it to the weight of the influence. Specifically, we can make the priority of influence between each two decision-making variables equal to the absolute value of the

---

[1] With the strongest influence.

[2] We use m to represent the number of issues and m to represent the number of alternatives for an issue.

© The Author(s), under exclusive license to Springer Nature Switzerland AG 2026   81
H. Luo, *Influence Models in Group Decision-Making*, Synthesis Lectures on Computer Science, https://doi.org/10.1007/978-3-032-01352-1_9

weight of influence between them. Therefore, for each preference/choice to be influenced, the influencing preference/choice that has the highest absolute value of the weight of influence on it has the highest priority of influence on it and can dominate the influence result. Intuitively, this rule for addressing multiple sources of influence across agents and issues can significantly reduce the complexity of the computation.

Moreover, although it is much more simplified compared with the rule of *multiple weighted influences*, in a sense, this rule of *one dominant influence* may be even closer to how people deal with multiple sources of influence (across agents and issues) in the real-world. For example, while facing a complicated case of multiple influences from many people making decisions on many issues, you will just follow (more broadly, be positively influenced by) your closet friend (or highest leader) or yourself on the most important issue for him or her or for yourself,[3] or just oppose (more broadly, be negatively influenced by) your most hated enemy on the most critical issue for him or her[4] (only if the absolute value of his or her weight of influence on this issue is the highest, regardless of whether the influence is positive or negative), rather than engage in a complex weighted deliberation[5] and computation.

Example 9.1 (A Display of How the Rule of One Dominant Influence Works When There are Multiple Weighted Influences across Agents and Issues) Assume the same case of multi-agent and multi-issue decision-making with a set of agents $\{1, 2, 3\}$ making decisions on a set of issues $\{X, Y, Z\}$,[6] each with three alternatives, as shown in Fig. 8.1. Although agent 3 faces complicated weighted influences across both multiple agents and multiple issues while making a decision on issue Z, he or she will just follow the influencing decision-making variable (i.e. the preference/choice of an agent on an issue) that has the highest priority of influence (i.e. the highest absolute value of the weight of influence), which is the preference/choice of himself or herself on former issue Y. This way to address multiple sources of influence is just like ignoring all other influences except the most predominant one, in mathematics, as if resetting the weights of all other influencing preferences/choices to zero (compared with Fig. 8.1), as shown in Fig. 9.1.

---

[3] Specifically, combining the closeness of amity and the importance of issue to determine the priority of influence of different decision-making variables, further, to determine which decision-making variable dominates the influence result.

[4] Specifically, combining the intensity of enmity and the importance of issue to determine the priority of influence of different decision-making variables, further, to determine which decision-making variable dominates the influence result.

[5] A elaborate consideration.

[6] Which can be the same, similar, or relevant/related issues at different times.

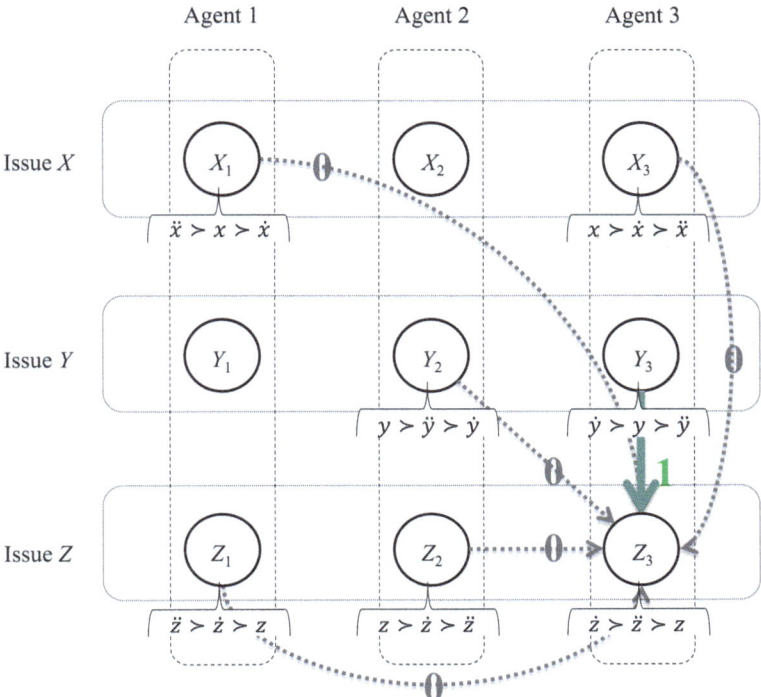

Fig. 9.1 A case of how the rule of one dominant influence works when there are multiple weighted influences across agents and issues. *Notes* Each node represents a decision-making variable (i.e. the preference/choice of an agent while making a decision on an issue). The link representing the one dominant influence that has the highest absolute value of the weight of influence is bold, and the links representing all other influences whose weight of influence has been reset to zero are dashed. Each influencing preference ordering is displayed (under its corresponding decision-making variable), including the original preference of the influenced agent

It should be noted that, since the latter self and the former self are represented by different decision-making variables (i.e. different nodes in the graph) in our framework, there will be no loop, even considering self influence.

Two Opposite Influences

10

Another rule we design for addressing multiple sources of influence across agents and issues is to assume that the preference/choice of an agent on an issue will be collectively affected by the strongest positively influencing preference/choice, the strongest negatively influencing preference/choice among all influencing preferences/choices of other agents on different issues, and his or her own former preference/choice with the strongest influence.

Two concepts, the *priority of positive influence* and the *priority of negative influence*, can be proposed to find or determine the strongest positively influencing preference/choice and the strongest negatively influencing preference/choice of other agents, respectively. Assume that there are $n \times m$ decision-making variables (corresponding to n agents making decisions on m issues) and that for each decision-making variable among them (corresponding to an agent making a decision on an issue), there is a priority of positive influence for each positive influence from other agents and a priority of negative influence for each negative influence from other agents while making decisions on different issues. The rule of *two opposite influences* assumes that each decision-making variable of an agent on an issue is collectively affected by the influencing decision-making variable of other agents that has the highest priority of positive influence on it, the influencing decision-making variable of other agents that has the highest priority of negative influence on it, and the decision-making variable of the same agent on former issues that has the strongest influence on it (i.e. on the current issue). Thus, we actually need to propose another concept of the *priority of self influence*. On the whole, there are three independent tracks for positive influences from others, negative influences from others, and self influences to compete (on priority).

Similarly, a simple and straightforward way to obtain or determine the priority of a positive influence, a negative influence or a self influence is to relate it to the weight of the influence. Specifically, we can make the priority of positive influence between two decision-making variables of one agent positively influencing another equal to the weight of influence between them, make the priority of negative influence between two decision-making variables of one agent negatively influencing another equal to the absolute value of the weight of influence between them, and make the priority of self influence between two decision-making variables of one single agent[1] equal to the absolute value of the weight of influence between them.[2] For the preference/choice of an agent on an issue to be influenced, the positively influencing preference/choice of other agents that has the highest weight of influence on it, the negatively influencing preference/choice of other agents that has the highest absolute value of the weight of influence on it, and his or her own preference/choice on former issues that has the highest absolute value of the weight of influence on it collectively determine the influence result (i.e. resulting preference/choice). The empathetic social choice [58], *social influence functions* and *matrix influence function* all can be extended to address such three sources of influence (a positive influence, a negative influence from others, and a self influence) across both agents and issues.

When we compare the three rules: *multiple weighted influences, one dominant influence,* and *two opposite influences*. The computational complexity of the rule of *multiple weighted influences* is the highest, which can be even unacceptable when there are too many agents making decisions on too many issues. Although the rule of *one dominant influence* is much more simple in the aspect of computation, it may be oversimplified compared to reality. Comparatively speaking, the rule of *two opposite influences* may achieve an appropriate balance and tradeoff between the computational complexity and explanatory power of the model describing the real-world influence.

This rule of *two opposite influences* may be very close to how people deal with multiple sources of influence (across agents and issues) in the real-world. For example, while facing a complicated case of multiple influences from many people making decisions on many issues, although you try to simplify the thinking process and focus, you will not ignore your closet friend's preference/choice on the most important issue for him or her,[3] your most hated enemy's preference/choice on the most critical issue for him or her,[4] and your own former preference/choice on the most important issue for yourself while computing and determining the influenced preference/choice.

---

[1] Which are one agent's preferences/choices on two issues.

[2] As mentioned earlier, self influences are usually positive but in extreme cases negative.

[3] Specifically, combining the closeness of amity and the importance of an issue for a friend to determine the priority of positive influences from others, in order to determine which positively influencing decision-making variable of other agents participates the computation of the influence result.

[4] Specifically, combining the intensity of enmity and the importance of an issue for an enemy to determine the priority of negative influences from others, in order to determine which negatively influencing decision-making variable of other agents participates the computation of the influence result.

Example 10.1 (A Display of How the Rule of Two Opposite Influences Works When There are Multiple Weighted Influences across Agents and Issues) Assume the same case of multi-agent and multi-issue decision-making with a set of agents {1, 2, 3} making decisions on a set of issues {X, Y, Z},[5] each with three alternatives, as shown in Fig. 8.1. Although agent 3 faces complicated weighted influences across both multiple agents and multiple issues while making a decision on issue Z, he or she will just take into account the positively influencing decision-making variable of other agents that has the highest weight of influence, the negatively influencing decision-making variable of other agents that has the highest absolute value of the weight of influence, and the decision-making variable of his or her own on former issues that has the highest absolute value of the weight of influence (on the current issue), which are the preference/choice of agent 1 on current issue Z, the preference/choice of agent 2 on former issue Y, and the preference/choice of himself or herself on former issue Y. This way to address multiple sources of influence is like ignoring all other influences except the strongest positive and negative influences from others and the strongest former self influence, in mathematics, as if resetting the weights of all other influencing preferences/choices to zero (compared with Fig. 8.1), as shown in Fig. 10.1.

It should be noted supplementally that, for the preference/choice of an agent on an issue to be influenced, if more than two positively influencing preferences/choices of other agents tie for the weight of influence on it, they can share a single highest weight of positive influence in the computation of the influence result. For example, if three positively influencing preferences/choices of other agents tie for the highest weight of 3, then all of them can participate in determining the resulting preference/choice, with each one using a weight of 1. The same processing can be used for negative influences in a tie and also self influences in a tie.

---

[5] Which can be the same, similar, or relevant/related issues at different times.

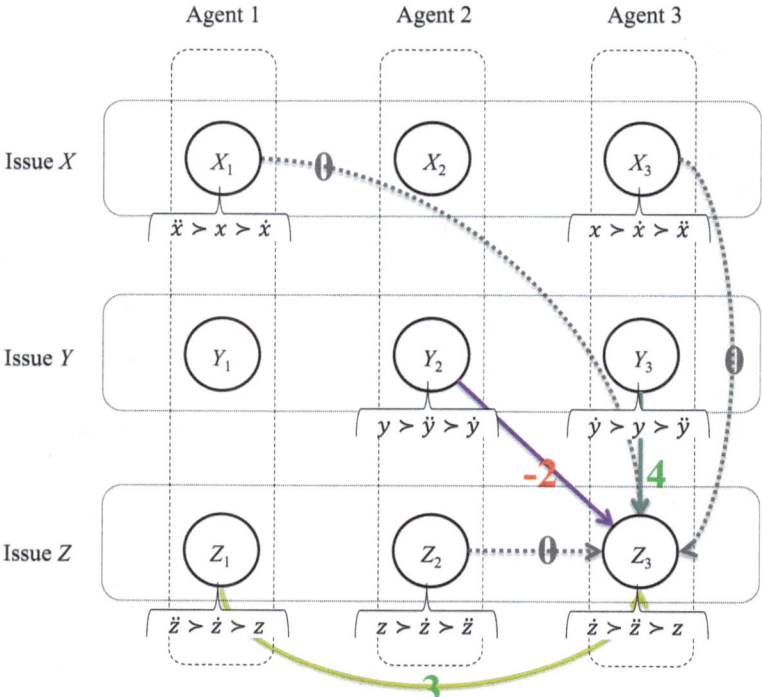

Fig. 10.1 A case of how the rule of two opposite influences works when there are multiple weighted influences across agents and issues. *Notes* Each node represents a decision-making variable (i.e. the preference/choice of an agent while making a decision on an issue). For the influenced decision-making variable at bottom right, the links representing the strongest positive and negative influences from others and the strongest former self influence that participate the computation of the influence result are bold, and the links representing all other influences whose weight of influence has been reset to zero are dashed. The weights of ignored influence are colored gray. Each influencing preference ordering is displayed (under its corresponding decision-making variable), including the original preference of the influenced agent

Discussion and Conclusion

We consider the scenario of group decisions where agents' decision-making behaviors and preferences can influence and be influenced by each other. To address multiple sources of influences among agents and obtain the choice or preference of an agent after being influenced by others, previous work discussed the situation of influence of one agent (at a time) on another agent or the simultaneous influence of more than one agent, mainly using cardinal approaches (such as employing utility value or score), but not ordinal approaches (such as employing preference ordering). In addition, the variation in the polarity (positive or negative) and strength (stronger or weaker) of different influences were sometimes ignored. In this book, we discuss how to address multiple sources of influence with varied strengths and opposite polarities not only in a nonordering approach, but also in an ordering-based approach by extending classical social choice functions to signed and weighted *social influence functions*, including *plurality influence function*, *majority influence function*, *Condorcet influence function*, and *Borda influence function*. In particular, we extend the KSB distance metric to a *matrix influence function*: we define the rule of how to transform each preference ordering into a matrix (named the ordering matrix) and set a measure to compute the distance between any two preference orderings (i.e. ordering matrices); then, the preference ordering[1] that has the smallest weighted sum of distances from all influencing agents' preferences orderings is the resulting preference for the influenced agent. Because the weight of influence can be positive or negative (e.g. from a friend or an enemy), it will play a role in finding the "closest" preference ordering while computing the distance from positively influencing preferences and finding the "farthest" preference ordering while computing the distance from negatively influencing preferences.

To address the simultaneous influence of more than one agent on another agent, previous work mainly discussed multiple influences in an individual way, assuming that all influ-

---

[1] At least theoretically existing.

© The Editor(s) (if applicable) and The Author(s), under exclusive license to Springer Nature Switzerland AG 2026
H. Luo, *Influence Models in Group Decision-Making*, Synthesis Lectures on Computer Science, https://doi.org/10.1007/978-3-032-01352-1

encing agents exert their own influences independently from each other and the resulting preference or choice of the influenced agent could be simply a linear weighted aggregation of all influencing agents' preferences or choices. Some previous work discussed the influence of coalitions of multiple agents. As for some influencing agents holding the same or similar beliefs, opinions or choices (such as an opinion alliance), an extra influencing effect besides the separate individual influences should be considered. However, another important influencing effect, which can be named *structural influence* has been largely ignored, in which the structure (i.e. influencing relationships among influencing agents)[2] should be addressed to determine the influence result. Actually, previous work has mainly perceived the structure (represented as links between agents in a social network) just as the path or channel of influence, while ignored the structure itself can also exert some extra influences as the source or origin of influence.[3] In fact, however, the influencing effect from structures on an agent is not easy to express and compute, as it involves two disparate categories of variables: the influencing subject is the influencing relationships among agents, while the influenced object is the preference or choice of a single agent. In this book, we demonstrate an elementary model of how to address the influence of structures on an agent, accompanied by the influence of coalitions of agents, on the basis of the influence of individual agents. Based on previous work and unique thoughts, we display a theoretical framework and systematic model of influence with three levels: the first level is the influence from independent agents (*individual influence*), the second level is the influence from coalitional agents (*coalitional influence*), and the third level is the influence from structured agents[4] (*structural influence*). The discussion of the *three levels of influence* in the context of group decision-making has both theoretical value and practical significance in a wide range of disciplines such as computer science, artificial intelligence (particularly multi-agent system), economics and management (particularly decision theory), and even politics and international relations (particularly international organization[5]).

In many circumstances, we need to discuss group decision-making not just regarding a singe issue or multiple issues independent to each other, but pertaining to multiple issues with combinatorial structures of dependencies among them, constituting a combinatorial and collective decision-making context. We should not only discuss the simultaneous influence of multiple agents on an agent while making a decision on a single issue or the simultaneous dependency on (influence by) multiple issues of an issue decided by a single agent, but also investigate the influence from multiple agents making decisions on multiple issues and address the conflicts among multiple sources of influence across both agents and issues

---

[2] In fact, in a combinatorial and collective decision-making context, the structure can indicate not only the influencing relationships among agents, but also the dependency relationships among issues.

[3] In another word, an agent can be influenced not only by the preferences or choices of the influencing agents (corresponding to individual influences), but also by the influencing relationships among the influencing agents.

[4] Which can also be understood as inter-influencing agents.

[5] Especially computational international organization theory [42–44].

with varied strengths (stronger or weaker) and opposite polarities (positive or negative). We provide a framework to address the influence across both multiple agents and multiple issues in combinatorial and collective decision-making with three rules of *multiple weighted influences*, *one dominant influence*, and *two opposite influences*. The first rule considers each source of influence according to its weight, ignoring none of multiple sources of influence; the second rule just focuses on the strongest influence, reduces the computational complexity significantly; the third rule takes into account the strongest positive and negative influences from others and the strongest self influence, reflecting the human psychology of simplifying problems but still seizing the main thread.

Future Work

There are some directions for future advancement:

- We have extended several classical social choice functions into sighed and weighted *social influence functions*, and demonstrated how to acquire the resulting preference or choice for an influenced agent when we choose one or another *social influence function*. It is common that different *social influence functions* lead to different results, just as different social choice functions lead to different results even under the same preference distribution. Therefore, how to evaluate different *social influence functions*, more specifically, how to find or formulate systematic benchmarks (criteria) to measure how good or how reasonable a *social influence function* is, or what properties a good *social influence function* should possess need to be discussed in the future, similar to the comparison and evaluation of *social choice functions* (group decision methods).

- While addressing the interplay between *structural influence* and *coalitional influence*, we only provide one elementary model by a probability-based approach (with two perspectives): decreasing the influenced one's weight of influence or increasing the uninfluenced one's weight of influence for *structural influence*; amplifying or reducing the decision-making probability referring to some classical social choice functions for *coalitional influence*. There may be many other meaningful analytical frameworks and mathematical models to be discussed about the *three levels of influence* in the future; indeed, as human minds are complicated, different people have various personalities, value systems and their own judgements, and even for a single person, his or her cognition will change under different environments, spaces, times, emotions, etc. Besides, an ordering-based approach for addressing the *three levels of influence* could be tried in future work.

- To address the multiple influences across both agents and issues, three framework models (rules) of *multiple weighted influences*, *one dominant influence*, and *two opposite influences* are proposed. For a multi-agent and multi-issue decision-making context where

H. Luo, *Influence Models in Group Decision-Making*, Synthesis Lectures on Computer Science, https://doi.org/10.1007/978-3-032-01352-1

issues are the same or similar to each other (belonging to the same or similar subjects, with the same set of alternatives) while happening at different times, *social influence functions*, *matrix influence function*, and the models of the *three levels of influence* we have built can be easily extended to adapt to a combinatorial and collective decision-making context. In future work, more analytic frameworks and mathematical models could be built and developed, particularly considering the case of multiple issues which are not similar to each other (i.e. with very different sets of alternatives), although such situation would be much more complicated.

Afterword

"If you are lucky enough to have lived in Paris as a young man, then wherever you go for the rest of your life, it stays with you, for Paris is a moveable feast."

Ernest Hemingway, A Moveable Feast

I would like to thank my supervisor, Professor Nicolas Maudet, for his guidance, support, and care during my stay in Paris. I would also like to thank my supervisors, Professor Qingguo Meng and Professor Yi Zhang, for their support at different stages of my academic journey. I would like to thank my family for their continued support and companionship.

Beijing, China   *Hang Luo*
May 2025

Glossary

Agent A general term that can represent a person or an artificial intelligence in nature and that can represent a decision-maker, a voter, a negotiator, or a game player, etc. in function.

Multiple sources of influence An agent is simultaneously influenced by more than one agent.

Three levels of influence Individual, coalitional and structural influence.

Influence across agents and issues Influence among multiple agents making decisions on different issues.

H. Luo, *Influence Models in Group Decision-Making*, Synthesis Lectures on Computer Science, https://doi.org/10.1007/978-3-032-01352-1

Author's Biography

Dr. Hang Luo is a tenured associate professor at Peking University in Beijing, China. He holds a Ph.D. in Computer Science from Universite Paris VI, Paris, France and a Ph.D. in Management from Tsinghua University, Beijing, China. His research interests include decision theory, network analysis, artificial intelligence (especially multi-agent system), and their applications. He has led three National Natural Science Foundation of China projects (72374010; 71804006; 7161101045).

© The Editor(s) (if applicable) and The Author(s), under exclusive license to Springer Nature Switzerland AG 2026
H. Luo, *Influence Models in Group Decision-Making*, Synthesis Lectures on Computer Science, https://doi.org/10.1007/978-3-032-01352-1

References

1. Antal, T., Krapivsky, P.L., Redner, S.: Social balance on networks: the dynamics of friendship and enmity. Physica D **224**(1), 130–136 (2006)
2. Bogart, K.P.: Preference structures I: distances between transitive preference relations. J. Math. Sociol. **3**(1), 49–67 (1973)
3. Boutilier, C., Brafman, R.I., Domshlak, C., Hoos, H.H., Poole, D.: CP-nets: a tool for representing and reasoning with conditional ceteris paribus preference statements. J. Artif. Intell. Res. **21**(1), 135–191 (2004)
4. Boutilier, C., Brafman, R.I., Domshlak, C., Hoos, H.H., Poole, D.: Preference-based constrained optimization with CP-Nets. Comput. Intell. **20**(2), 137–157 (2004)
5. Brachman, R.J., Levesque, H.J.: Machines Like Us: Toward AI with Common Sense. The MIT Press, Cambridge (2022)
6. Brandes, U.: On variants of shortest-path betweenness centrality and their generic computation. Soc. Netw. **30**(2), 136–145 (2008)
7. Brandes, U., Erlebach, T. (eds.): Network Analysis: Methodological Foundations. LNCS, vol. 3418. Springer, Berlin, Heidelberg (2005)
8. Brandes, U., Fleischer, D.: Centrality measures based on current flow. In: Proceedings of the 22nd Annual Symposium on Theoretical Aspects of Computer Science. LNCS, vol. 3404, pp. 533–544 (2005)
9. Brandt, F., Conitzer, V., Endriss, U.: Computational social choice. In: Weiss, G. (ed.) Multiagent Systems, pp. 213–283. The MIT Press, Cambridge (2013)
10. Brandt, F., Conitzer, V., Endriss, U., Lang, J., Procaccia, A.D. (eds.): Handbook of Computational Social Choice. Cambridge University Press, Cambridge (2016)
11. Capuano, N., Chiclana, F., Fujita, H., Herrera-Viedma, E., Loia, V.: Fuzzy group decision making with incomplete information guided by social influence. IEEE Trans. Fuzzy Syst. **26**(3), 1704–1718 (2018)
12. Chevaleyre, Y., Endriss, U., Lang, J., Maudet, N.: A short introduction to computational social choice. In: Proceedings of the 33rd Conference on Current Trends in Theory and Practice of Computer Science. LNCS, vol. 4362, pp. 51–69 (2007)
13. Degroot, M.H.: Reaching a consensus. J. Am. Stat. Assoc. **69**(345), 118–121 (1974)
14. Demarzo, P.M., Vayanos, D., Zwiebel, J.: Persuasion bias, social influence, and unidimensional opinions. Q. J. Econ. **118**(3), 909–968 (2003)

© The Editor(s) (if applicable) and The Author(s), under exclusive license to Springer Nature Switzerland AG 2026

H. Luo, *Influence Models in Group Decision-Making*, Synthesis Lectures on Computer Science, https://doi.org/10.1007/978-3-032-01352-1

15. Freeman, L.C.: Centrality in social networks conceptual clarification. Soc. Netw. **1**(3), 215–239 (1978)
16. Freeman, L.C., Borgatti, S.P., White, D.R.: Centrality in valued graphs: a measure of betweenness based on network flow. Soc. Netw. **13**(2), 141–154 (1991)
17. Friedkin, N.E., Johnsen, E.C.: Social influence and opinions. J. Math. Sociol. **15**(3–4), 193–206 (1990)
18. Friedkin, N.E., Johnsen, E.C.: Social positions in influence networks. Soc. Netw. **19**(3), 209–222 (1997)
19. Golub, B., Jackson, M.O.: Naive learning in social networks and the wisdom of crowds. Am. Econ. J. Microecon. **2**(1), 112–149 (2010)
20. Grabisch, M., Rusinowska, A.: Measuring influence in command games. Soc. Choice Welfare **33**(2), 177–209 (2009)
21. Grabisch, M., Rusinowska, A.: A model of influence in a social network. Theor. Decis. **69**(1), 69–96 (2010)
22. Grabisch, M., Rusinowska, A.: A model of influence with an ordered set of possible actions. Theor. Decis. **69**(4), 635–656 (2010)
23. Grabisch, M., Rusinowska, A.: Influence functions, followers and command games. Games Econ. Behav. **72**(1), 123–138 (2011)
24. Grabisch, M., Rusinowska, A.: A model of influence based on aggregation function. Math. Soc. Sci. **66**(3), 316–330 (2013)
25. Grandi, U., Lorini, E., Perrussel, L.: Propositional opinion diffusion. In: Proceedings of the 14th International Conference on Autonomous Agents and Multiagent Systems, pp. 989–997 (2015)
26. Grandi, U., Luo, H., Maudet, N., Rossi, F.: Aggregating CP-Nets with unfeasible outcomes. In: Proceedings of the 20th International Conference on Principles and Practice of Constraint Programming. LNCS, vol. 4362, pp. 366–381 (2014)
27. Granovetter, M.: The strength of weak ties: a network theory revisited. Soc. Theory **1**(6), 201–233 (1983)
28. Granovetter, M.: The impact of social structure on economic outcomes. J. Econ. Perspect. **19**(1), 33–50 (2005)
29. Granovetter, M.S.: The strength of weak ties. Am. J. Sociol. **78**(6), 1360–1380 (1973)
30. Hao, F., Ryan, P.Y. (eds.): Real-World Electronic Voting: Design, Analysis and Deployment. CRC Press, Boca Raton (2016)
31. Harary, F.: On the notion of balance of a signed graph. Mich. Math. J. **2**(2), 143–146 (1953)
32. Harrison, L., Bag, S., Luo, H., Hao, F.: Vericondor: end-to-end verifiable condorcet voting with support for strict preference and indifference. Digit. Threats **5**(4), 1–30 (2024)
33. Heider, F.: Attitudes and cognitive organization. J. Psychol. **21**(1), 107–112 (1946)
34. Hoede, C., Bakker, R.: A theory of decisional power. J. Math. Sociol. **8**(2), 309–322 (1982)
35. Hu, X., Shapley, L.S.: On authority distributions in organizations: controls. Games Econ. Behav. **45**(1), 153–170 (2003)
36. Hu, X., Shapley, L.S.: On authority distributions in organizations: equilibrium. Games Econ. Behav. **45**(1), 132–152 (2003)
37. Jackson, M.O.: Social and Economic Networks. Princeton University Press, Princeton (2008)
38. Joy, G.: Ballots in the belfry: Lewis Carroll and voting fairness (2002). https://nmtx.org/details/article/ballots-in-the-belfry-lewis-carroll-and-voting-fairness
39. Kemeny, J.G., Snell, J.L.: Mathematical Models in the Social Sciences. The MIT Press, Cambridge (1972)
40. Krackhardt, D.: The strength of strong ties: the importance of Philos in organizations. In: Rob Cross, A.P., Sasson, L. (eds.) Networks in the Knowledge Economics, pp. 82–105. Oxford University Press, New York (2003)

41. Luo, H., Guo, Z., Zhang, Y.: Value analysis of mobile government. Inf. Doc. Serv. **4**, 36–40 (2010). (In Chinese)
42. Luo, H.: Agent-based computing experiments of international organization decision-making: illustrated by the example of the evolution of the membership composition and the change of the decision-making system of the EU. World Econ. Polit. **7**, 120–155 (2020). (In Chinese)
43. Luo, H., Meng, Q.: Multi-agent simulation of the UN SC reform and great power games. World Econ. Polit. **6**, 136–155 (2013). (In Chinese)
44. Luo, H.: Computational International Organization Theory. People's Press, Beijing (2025). (In Chinese)
45. Luo, H., Wang, Z., Yang, S., Yang, H., Gong, Y.: Influence among preferences and its transformation to behaviors in groups. In: Proceedings of the 20th International Conference on Group Decision and Negotiation. LNBIP, vol. 388, pp. 104–119 (2020)
46. Luo, H., Yang, L.: Equality and equity in emerging multilateral financial institutions: the case of the BRICS institutions. Glob. Policy **12**(4), 482–508 (2021)
47. Luo, H., Yang, L., Houshmand, K.: Power structure dynamics in growing multilateral development banks: the case of the Asian Infrastructure Investment Bank. Glob. Policy **12**(1), 24–39 (2021)
48. Maran, A., Maudet, N., Pini, M.S., Rossi, F., Venable, K.B.: A framework for aggregating influenced CP-nets and its resistance to bribery. In: Proceedings of the Twenty-Seventh AAAI Conference on Artificial Intelligence, pp. 668–674 (2013)
49. Maudet, N., Pini, M.S., Venable, K.B., Rossi, F.: Influence and aggregation of preferences over combinatorial domains. In: Proceedings of the 11th International Conference on Autonomous Agents and Multiagent Systems, pp. 1313–1314 (2012)
50. NeuroLaunch: Peer pressure and mental health: exploring the profound impact on well-being (2025). https://neurolaunch.com/how-does-peer-pressure-affect-mental-health/
51. Newman, M.: Networks, 2nd edn. Oxford University Press, New York (2018)
52. Newman, M.E.J.: Scientific collaboration networks I network construction and fundamental results. Phys. Rev. E **64**, 016131 (2001)
53. Newman, M.E.J.: Scientific collaboration networks II shortest paths, weighted networks, and centrality. Phys. Rev. E **64**, 016132 (2001)
54. Newman, M.E.J.: A measure of betweenness centrality based on random walks. Soc. Netw. **27**(1), 39–54 (2005)
55. Pérez, L.G., Mata, F., Chiclana, F., Gang, K., Herrera-Viedma, E.: Modelling influence in group decision making. Soft. Comput. **20**(4), 1653–1665 (2016)
56. Purrington, K., Durfee, E.H.: Making social choices from individuals' CP-nets. In: Proceedings of the 6th International Conference on Autonomous Agents and Multiagent Systems, pp. 1–3 (2007)
57. Rossi, F., Venable, K.B., Walsh, T.: A Short Introduction to Preferences: Between Artificial Intelligence and Social Choice (Synthesis Lectures on Artificial Intelligence and Machine Learning). Springer, Cham (2022)
58. Salehi-Abari, A., Boutilier, C.: Empathetic social choice on social networks. In: Proceedings of the 13th International Conference on Autonomous Agents and Multiagent Systems, pp. 693–700 (2014)
59. Stone, P.: Intelligent Autonomous Robotics (Synthesis Lectures on Artificial Intelligence and Machine Learning). Springer, Cham (2022)
60. Terzi, E., Winkler, M.: A spectral algorithm for computing social balance. In: Proceedings of the 8th International Workshop on Algorithms and Models for the Web-Graph. LNCS, vol. 6732, pp. 1–13 (2011)
61. Wasserman, S., Faust, K.: Social Network Analysis: Methods and Applications. Cambridge University Press, Cambridge (1994)

62. Wicker, A.W., Doyle, J.: Interest-matching comparisons using CP-nets. In: Proceedings of the Twenty-Second AAAI Conference on Artificial Intelligence, pp. 1914–1915 (2007)
63. Wicker, A.W., Doyle, J.: Comparing preferences expressed by CP-networks. In: Proceedings of the AAAI Workshop on Advances in Preference Handling, pp. 128–133 (2008)